DATE DUE			
OCT 29 '85			

TECHNIQUES FOR MANAGING TECHNOLOGICAL INNOVATION
Overcoming Process Barriers

TECHNIQUES FOR MANAGING TECHNOLOGICAL INNOVATION
Overcoming Process Barriers

By

CAROL WALCOFF
ROBERT P. OUELLETTE
PAUL N. CHEREMISINOFF

ANN ARBOR SCIENCE
THE BUTTERWORTH GROUP

PREFACE

This book presents a study that investigates management techniques used by U.S. industry and government to facilitate the technological innovation process. The study discusses how selected management techniques can help overcome technical, organizational, governmental, financial and marketing barriers in the innovation process. Descriptions of 21 management techniques are included.

Management techniques and candidate organizations using these techniques are presented with information on their relative abilities to stimulate development and commercialization of new technological markets. The management techniques selected for presentation are currently employed as a means of managing technological innovation. The processes are defined by which industry generates new and improved products and methods of production.

Activities are included that range from idea generation, research, development and commercialization to diffusion through the economy. This book should be useful to engineers and scientists involved in production, product and management functions, as well as managers, executives, sales personnel and all those who are part of their organization's decision-making process. Technological innovation—bringing new products and processes into the marketplace—is made up of many small steps, from the thought of a new concept to its final sale and use.

In addition to the rationale and background involved in the management techniques for technological innovation, this book presents "case histories" of major organizations, covering a broad spectrum of styles and approaches. The reader will find these presentations valuable for study and reference. The approach here is not one of the traditional "cookbook," but, we hope, a text the reader can apply in the thought process toward his or her own business activities.

Electricité de France is acknowledged for its support and encouragement, which made this book possible.

Carol Walcoff
Robert P. Ouellette
Paul N. Cheremisinoff

Walcoff   **Ouellette**   **Cheremisinoff**

Carol Walcoff has been a member of the technical staff at the MITRE Corporation for more than six years. Her work has focused on information resource management, organizational and program development, policy research, and system modeling and evaluation. She has conducted work for private clients and governmental sponsors in the fields of health and social services delivery, telecommunications system design, occupational health impact evaluation, international environmental programs, and the management of technological innovation. She received her MBA from Southern Illinois University and BA from the University of Wisconsin.

Robert P. Ouellette is Technical Director of the Environment Division of the MITRE Corporation. Dr. Ouellette has been associated with MITRE in varying capacities since 1969. He is a graduate of the University of Montreal and received his PhD from the University of Ottawa. A member of the American Statistical Association, Biometrics Society, Atomic Industrial Forum and the National Science Foundation Technical Advisory Panel on Hazardous Substances, Dr. Ouellette has published numerous technical and scientific papers and books on a wide variety of subjects. He is co-author and co-editor of the comprehensive *Electrotechnology* survey series, as well as other books published by Ann Arbor Science.

Paul N. Cheremisinoff is Associate Professor at the New Jersey Institute of Technology, Newark. A registered Professional Engineer and consulting engineer, he has been a consultant on environment/energy/resources projects for the MITRE Corporation. An internationally known scholar and researcher, he is author/editor of many Ann Arbor Science publications in engineering/energy/environmental control, including *Pollution Engineering Practice Handbook, Carbon Adsorption Handbook* and *Environmental Impact Data Book*.

CONTENTS

ix

FIGURES AND TABLES

FIGURES

TABLES

xiv

CHAPTER 1

INTRODUCTION

This book presents a review of the applications of management techniques currently being used by U.S. industry and government to encourage or facilitate technological innovation. Management techniques and candidate organizations using the techniques were identified during the first phase of the study. In the second phase, users of the management techniques were interviewed to obtain information regarding the relative effectiveness of the management techniques to stimulate the development and commercialization of new technological products and processes. The group of management techniques included do not necessarily represent the most successful means of managing technological innovation but were selected because they are in current use. Further, information regarding their application was accessible to investigation at this time.

Technological innovation has been defined as "the process by which industry generates new and improved products and production processes. It includes activities ranging from the generation of an idea, research, development and commercialization to the diffusion throughout the economy of new and improved products, processes and services" [1]. Innovation in the academic sense generally refers to the process of creating a new product or process; however, for this report, that definition is extended to include those existing products and processes for which new users or markets may be defined. Technological innovation occurs as a result of either technology push or market pull. Technology push for innovation happens when innovators seek commercially viable uses for new or existing technology. With market pull, innovation results from the recognition of an existing problem or need by the innovator, as well as the application of technical and capital resources to develop alternative technological solutions.

The process of technological innovation—bringing new products and processes into the market place—actually consists of many small steps from "thinking" of a new concept to "selling" it. The steps do not always follow in a consecutive stepwise fashion for all industries, nor are all steps necessarily required in all cases. A diagram of some of the major steps in the technological innovation process is presented in Figure 1. These steps have been grouped into three major phases. Elements of this innovation process model were drawn from the literature [2,3] and are used in this study to describe the applications of the management techniques. The first phase consists of setting broad goals regarding innovation policy, identifying the general types of new products and processes to be developed, developing innovative ideas, selecting innovation projects and conducting preliminary project planning. The second phase includes research and development (R&D) on both production methods and the products. The third phase consists of marketing or diffusion of the new product or process.

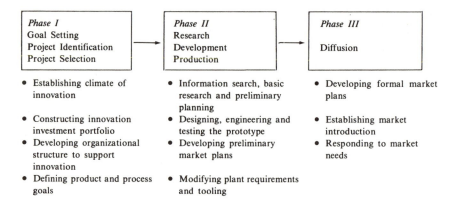

Figure 1.   Three phases of the innovation process.

Corporate decision-makers elect to invest in R&D supporting technological innovation for several reasons. They may wish to identify new products or processes that will permit the firm to remain competitive by providing new ways to reduce costs and improve quality, or they may wish to expand existing market areas. Businesses that tend to support ongoing investment in innovation also believe it may be advantageous to maintain an R&D effort as a resource for production-related problem-solving.

There are, however, a number of barriers within the innovation process that inhibit the development of new products and processes. These barriers may be manifested in the lack of technical or financial resources required to support innovation or may be defined as structural deficiencies within an organization inhibiting the flow of innovative ideas. These barriers can retard or stop the technological innovation process within an organization. Recognizing these barriers as impediments to the innovation process, business managers in U.S. industry and government have developed management techniques to overcome them and increase the probability that the development of new products and processes will result in commercial success.

This book provides a summary of 21 management techniques selected for review and used to effect innovation process barriers. A brief description of each technique is presented in Table I. Although these techniques address problems that occur during the innovation process, they were not necessarily developed solely for that function. For example, some address general management problems associated with personnel, finance, facilities, etc. The information gathered during interviews with users of the management techniques is summarized in Chapter 3. A series of tables are presented in that chapter to assist the reader in analyzing problems associated with managing the innovation process and to identify keys to potential solutions based on the experience of other organizations and industries.

Table I. Brief Descriptions of the Management Technique

Management Technique	Description
1. Idea Generation	New product and process ideas are generated through both structured and unstructured discussions involving technical staff and management.
2. Innovation Incentives	A financial reward and recognition program is established to encourage technical professionals to stay within the firm and to increase the rate and level of innovation through the development of alternative career ladders designed to maintain professional interest and challenges.
3. Innovation Training	Employees throughout the organization participate in organizational development programs that encourage greater receptivity in nurturing innovative ideas, particularly through group dynamics.
4. Internal Venture Management	Organizations isolate the financial risks and rewards associated with management of the innovation process, establishing a profit center along process lines.

Table I, continued

Management Technique	Description
5. Production Champion	An individual is identified to act as political surrogate and manager responsible for the development of a new product or process.
6. Project Performance Measurement	Guidelines are established by innovation project management and staff to evaluate the effectiveness of project activities and their outcomes.
7. Quality Circles	Voluntarily formed groups of nontechnical and technical staff meet to identify problems within the organization's operation that require innovative solutions.
8. Research Planning Frame	A structured corporate research planning process is utilized to ensure participation of experts in the long-term identification of product/process goals.
9. Temporary Groups	Individuals representing several levels of management form committees as a temporary work assignment to apply problem-solving techniques to develop new product/process ideas.
10. External Venture Management	Intensive capital, marketing and management resources outside the organization are utilized to initiate new product development or to assist in mature firm development and growth.
11. Franchising	A legal relationship is established between a corporation and distributors of its products and services to facilitate the transfer of innovative product/processes.
12. Licensing	The developer of an innovative product or process permits its use and further development but maintains legal control and receives payment for these license privileges.
13. Middleman Broker	Universities' services and resources are matched with industry's technical innovation needs through the coordinating efforts of a not-for-profit technical organization.
14. Personnel Transfers	Technical staff are encouraged to transfer both within the organization as well as among organizations to stimulate the transfer of technological innovation.
15. University-Industry Linkages	Business management services are provided to small business owners and operators and individual inventors to assist in the development of strategies for managing new product development.
16. Industrial Applications Centers	Technical personnel and information resources have been organized in a center to assist in the public dissemination of new technology developed for space research.
17. Innovation Centers	University-based centers provide startup funding and technical and management support to assist individual inventors and innovators in new products/process development.
18. National Technology Foundation	Dispersed professional, engineering and other technical resources are consolidated under federal direction to address national problems potentially requiring technological solutions.
19. Regulations	Federal legislation and guidelines may stimulate the dis-

Table I, continued

Management Technique	Description
	semination and use of new products and processes by creating an incentives structures for the users.
20. Technology Transfer from Laboratories	Industry representatives are involved in the development of new products and processes at government laboratories to ensure full understanding of the innovation concept and effective commercialization.
21. Subsidies	Federal funding of industries involved in technological innovation requiring high R&D investments ensures technological advancement and reduction of risk borne by the innovators.

CHAPTER 2

METHODOLOGY

STUDY APPROACH

The goal of this study is to identify and describe management techniques used to facilitate the innovation process and the exploitation of new products and processes. The study objectives defined to achieve this goal are (1) to identify and characterize the aspects of management techniques that facilitate the diffusion of innovation; (2) to gain a better understanding of the barriers to the diffusion of innovation through the analysis of management techniques applied in different stages of the innovation process; and (3) to obtain information regarding the successful or unsuccessful use of management techniques.

Three major tasks were pursued. The first was to identify potentially useful management techniques and candidate organizations to be included in the study. The second was to obtain additional information from individuals in government and business who use the selected management techniques. The third objective was to review the information collection and prepare a summary and analysis of the interview results.

SELECTION OF MANAGEMENT TECHNIQUES AND CANDIDATE ORGANIZATIONS

The management techniques were selected by reviewing the literature (1975–1981) to identify a group of techniques that would address major barriers to the innovation process. The information search consisted of a formal literature search and interviews with individuals involved in the

7

study of innovation management. The literature search focused on innovation theory and technological change rather than research regarding management theory. More than 200 documents were reviewed, including bibliographies that have been prepared on related topics; government hearings and reports concerning small business programs, tax incentives or subsidies and innovation policies; published case studies; and journal articles (1975–1980). The *Business Periodicals Index* and the *National Technical Information Service Directory* were the major reference sources.

To assist in the identification of management techniques, interviews were conducted with those individuals in the U.S. government responsible for the development of national policies and programs to stimulate technological innovation. These interviews provided information on successful techniques that have been studied by different government organizations and suggestions concerning followup contacts in specific organizations.

Sets of criteria were developed to select (1) management techniques and (2) the candidate organizations to be included in the case history analysis. Both sets, as described below, were based heavily on the expressed recommendations of Electricité de France (EDF).

Criteria for Selection of Management Techniques

1. The techniques must be transferable to the EDF organization and the French industry setting. The techniques should not be limited by product, setting, industry, market or government regulation. The technique must be applicable to more than one type of organization or industry.

2. The techniques must address major barriers occurring during the innovation process, including those that may be internal or external to the organization. These barriers may reflect philosophical, structural or resource limitations or refer to the lack of resources, such as manpower (which includes technical expertise or knowledge), capital or raw materials.

Criteria for Selection of Candidate Organization

The selection of candidate organizations was conducted concurrently with that of the management techniques. Two general criteria were used in this process.

1. The list of candidate organizations included those that have successfully, and unsuccessfully, applied the selected management techniques.

2. The list of candidate organizations was representative of different organizational structures and product orientations. It included the following:

- organizations that develop advanced and rapidly changing technology;
- stable regulated organizations with significant research programs;
- organizations that sponsor large, diversified research programs; and
- organizations greatly affected by government policy and regulation.

Application of the selection criterial was accomplished through the coordinated review of recommended management techniques and candidate organization.

For discussion, the selected group of management techniques may be said to consist of three sets: (1) those applied within the organization; (2) those that occur as a result of management decisions of two or more organizations; and (3) those that are initiated by the government. Figure 2 presents a graphic illustration of the application of these management techniques, the management characteristics associated with each group and a list of the selected management techniques.

DATA COLLECTION

Interview guides were prepared to assist in the data correction efforts prepared during both initial contacts with the organizations' representatives and site visit interviews. The first was structured to obtain information on the organization, the industry of which it is a member and potential contacts to be made during the site visit to obtain information on the use of the subject management technique. The second interview guide assisted the interviewer in obtaining information to be included in descriptions of the applications of management techniques.

Prior to interviews with site contacts, research was conducted to obtain more information on the management techniques to be studied and the organizations to be visited. Site contacts were telephoned prior to the site visits to obtain additional information on the organizations' structures and applications of the management techniques using the first interview outline described. The interviewer provided the site contacts with a detailed description of the information desired. An effort was made during this interview to determine whether it was necessary to meet with other individuals in the organization during the site visit to obtain the information desired.

Background data concerning the candidate organizations were ob-

Internal Management Techniques are initiated and applied within the business.

1. Idea Generation
2. Innovation Incentives
3. Innovation Training
4. Internal Venture Management
5. Product Champion
6. Project Performance Measurement
7. Quality Circles
8. Research Planning Frame
9. Temporary Groups

External Management Techniques are initiated by one business to affect another usually by providing technical or financial resources

10. External Venture Management
11. Franchising
12. Licensing
13. Middleman/Broker
14. Personnel Transfers
15. University-Industry Linkages

Government-Sponsored Management Techniques are initiated by the government to impact technological innovations developed by the individual, business, or other organizations.

16. Industrial Application Centers
17. Innovation Centers
18. National Technology Foundation
19. Regulations
20. Technology Transfer from Laboratories
21. Subsidies

Figure 2.   Classification of management techniques.

tained primarily by reviewing business periodicals, corporate annual reports, and business and government data summaries. During this phase, the literature survey was focused on the selected management techniques. Specific information regarding use of the techniques was obtained by reviewing the current literature and interviewing management representatives of organizations employing the techniques. The representatives were managers or individuals who had been involved in the development and application of the technique (consultants, academics). By this approach, an understanding was gained of each technique's problems, successes and other potential applications.

An informal interview guide was prepared for the interview sessions, consisting of questions about the characteristics of the organization and the methods it used to select, implement and evaluate the management technique. The one- to two-hour interviews generally were conducted with senior corporate staff or project leaders at the site of the candidate organization. When possible, other individuals within the organization were contacted to obtain additional perspectives on the management technique.

ANALYSIS OF THE MANAGEMENT TECHNIQUES

Review of the applications of the selected management techniques could have been structured in a number of ways, such as focusing on the classification of resources required to implement the technique or by characterizing the relative size and structures of the organizations adopting the techniques. For discussion, the innovation process barriers affected by the management techniques were selected as a means for structuring the review. The set of barriers was based on a review of studies that have utilized barriers as an analytical tool to provide a basis for describing approaches to managing innovation. Five broad groups of barriers emerged from this review [3,4]. They are as follows:

1. **Technical Barriers**—pertain to the lack, or inefficient use, of technology-related resources, including expertise (manpower), state-of-the-art knowledge, or raw materials. Sample specific technical barriers include:
   - Lack of technical information. This inhibits the process by curbing all phases of the process limiting the scope of innovation ideas.
   - Lack of support for collateral invention. Individuals are not encouraged to participate in creative interaction with peers to stimulate innovative interchanges.
   - Lack of other technical resources. This barrier involved the lack of appropriate personnel to permit the development of new processes and products or may refer to the lack of technology required to develop or manufacture new products and processes.

2. **Organizational Barriers**—occur within the organization or firm and usually involve the communications among, or organizational structure of, the management and innovation process participants. Some of the specific organizational barriers that impart the innovation process are as follows
   - Lack of support from top management. Without the support of top management, the innovation process may falter for lack of adequate goal setting and financial backing.
   - Environment not supportive of innovators. If the innovators within the organization do not sense an environment that nourishes and nurtures creative thought, the developmental process may be slowed or stopped.
   - Insufficient communications channels. Those linkages are vital to the innovation process to ensure that the corporate philosophy is understood by all levels of the organization and to maximize innovation project operation.
   - Lack of adequate professional recognition and reward. Technical specialists, as well as managers of innovation, must be compensated appropriately to encourage personal and professional growth.
   - Ineffective project management. If innovation projects are not managed efficiently, the process may be affected.
3. **Governmental Barriers**—may exist in the form of regulations (anticipated or real), a tax structure or subsidies. Two are as follows:
   - Uncertainty concerning regulatory policies. Innovators may be discouraged from investing in the development of new products and processes if government regulatory policies hamper their diffusion.
   - Inadequate patent protection. By not protecting the rights of the individual inventor or inventor's organization, the new product or process may be "lost" and not successfully developed.
4. **Financial Barriers**—affect the flow of capital into innovation projects that appear technically feasible. Financial barriers include the following:
   - High financial risk associated with a particular process phase. Each of the three major phases of the innovation process has different levels of risk associated with the development of new products. These levels vary from industry to industry and may be measured and determined unacceptable.
   - Inability to use forecasts and estimates. An organization must take advantage of forecasting methods to assess the relative worth of different innovation paths. If the organization cannot do this due to the unavailability of such estimates or the inability to use them accurately, it will falter in the planning or diffusion stage.
   - Lack of financial resources assigned to innovation. Innovation projects require planned financial support throughout all phases of the process to ensure new product marketability.
5. **Marketing Factors**—impact the diffusion of new technology or the process of perceiving and acquiring new markets for technological innovation. Specific marketing barriers relate to information and management need:
   - Lack of information concerning characteristics of potential markets. This barrier actually may occur early in the product planning phase but may have impacts throughout the process. If a market need is not determined accurately, the diffusion of the new technology cannot be predicted.
   - Inability to coordinate technical and marketing goals. Although market needs may be assessed accurately, it is important that they be translated into viable technical and marketing goals and objectives.

The information collected during the interviews was used to categorize and compare management techniques. Statistical analyses of the application of associated behavior patterns were not possible due to the diversity of organizations and situations surveyed. Rather, the information is presented as a profile of selected management methods used to overcome barriers associated with technological innovation.

CHAPTER 3

FINDINGS

OVERVIEW

This chapter highlights the applications of the management techniques. The tables that follow and the descriptions of the management technique applications presented in Chapter 4 may be used to identify potential solutions to innovation management problems. First, the techniques are discussed in summary form to compare the organizations and general characteristics of the techniques. Then, in more detail, a series of tables illustrates barriers affected throughout the innovation process.

The reader may make the most effective use of this information by defining a problem in terms of both the class of barriers—technical, organizational, governmental, financial or marketing—and of the phase of the innovation process during which the problem or barrier occurs. The next step is to identify a management technique(s) currently in use that has affected similar problems or barriers. The tables in this chapter are divided first by phase, then are presented to correspond with the classes of barriers. To identify specific management techniques, select the appropriate phase and barrier class; the management techniques can be compared by reviewing their effects on specific problems as presented in Tables V–XVIII.

A summary of some of the major characteristics of the management techniques and organizations surveyed is presented in Table II. It points out key issues influencing the selection of a management technique and its evaluation by the organization. During the interview, it was discovered that some selected management techniques were used by more than one candidate organization. Time did not permit the collection of data for multiple techniques; however, those organizations using each technique are listed in the second column of Table II. The organization interviewed about the management technique is listed first. As indicated by the 1980 revenues and employment figures, most commercial for-profit firms are relatively large ($1.5 million to $43.5 billion in 1980 revenues). The last column in Table II describes the organizations' methods of evaluating the management techniques. In many cases, they are still being evaluated to assess their success in terms of the indicators listed in this column.

As indicated earlier, the innovation process consists of a broad range of activities involving virtually all aspects of an organization's operations. The selected management techniques may only be applicable to certain phases of the process, however, depending on the barriers to be addressed. Table III indicates the phases of the innovation process during which the management techniques are applied, and Table IV summarizes the groups of barriers affected by the management techniques.

MANAGEMENT TECHNIQUES AFFECTING BARRIERS WITHIN THE FIRST PHASE OF THE INNOVATION PROCESS

The first phase of the innovation process encompasses activities regarding the definition of an overall corporate innovation goal structure, as well as the selection and funding of individual projects. Four classes of barriers were affected by management techniques during this first phase—technical, organizational, financial and marketing.

Technical barriers may be described as the lack of technical information, such as manpower or technological data within the firm, to stimulate the creation of new product or process ideas. The techniques that address technical barriers within the first phase, such as idea generation, the research planning frame and industrial applications centers, provide unique methods of tapping potentially creative technical resources to support goal setting and new product planning.

Organizational barriers are important as they may inhibit the flow of new product ideas through the corporate information channels or dampen the innovative spirit. Internal venture management and temporary groups are techniques that address the lack of support from top management by involving it in the innovation planning process. Innovation training and idea generation have been used to overcome environmental barriers by creating structured relationships among innovators in the development of new technologies. Innovation incentives have been used to create a communications infrastructure and a professional recognition and rewards program to strengthen organizational linkages

Table II.   Summary of Surveyed Management Techniques and Organizations

Management Technique	Organization[a]	Organization's Products/Services	Organization's 1980 Revenues and Employment	Focus of the Management Technique	Factors Contributing to the Success of the Management Technique	Method of Evaluating the Management Technique
1. Idea Generation	Ford Motor Company (Bell Laboratories) (The MITRE Corporation) (University of Utah) (National Technology Foundation)	• Automobiles, Trucks, tractors, and implements	• $43.5 billion • 500,000 employees	• To conduct long range planning of innovation within a structured environment.	• Involvement of leading experts/innovators in the idea generating process	• Measurement of profits resulting from selection and development of ideas
2. Innovation Incentives	International Harvester Company (Xerox Corporation) (Bell Laboratories) (University of Utah) (Utility Firms)	• Trucks, agricultural and industrial equipment, construction equipment, recreational vehicles, lawn and garden equipment, engines, and steel	• $8.3 billion • 90,000 employees	• To encourage technical professionals to stay with the firm and increase level and rate of innovation.	• Establishment of alternative career ladders to maintain professional interest and challenges	• Measurement of numbers of new inventors, patents and awards
3. Innovation Training	BF Goodrich Company (Xerox Corporation) (Bell Laboratories) (University of Wisconsin) (University of Utah) (Utility Firms)	• Tires and tubes, industrial rubber products, aircraft wheels and brakes and aircraft parts, chemicals, plastics, synthethic rubbers, and petrochemicals	• $3.2 billion • 41,000 employees	• To encourage staff to be more receptive in nurturing innovative ideas.	• Forced involvement of individuals trained in group dynamics in the management of innovation	• Measurement of the number of innovative ideas suggested and developed with and without the training experience
4. Internal Venture Management	3M Company (International Harvester Company) (BF Goodrich Company) (International Telephone and Telegraph Corporation) (McDonald's Corporation) (Bell Laboratories) (University of Utah)	• Coated and bonded adhesive materials	• $5 billion • 83,000 employees	• To concentrate and isolate the risks (and rewards) associated with the innovation process from the firm's normal operations.	• The individuals involved in the project have greater incentives to succeed as risks and rewards are parallel	• Measurement of profits associated with sales of new product or process

Table II, continued

Management Technique	Organization[a]	Organization's Products/Services	Organization's 1980 Revenues and Employment	Focus of the Management Technique	Factors Contributing to the Success of the Management Technique	Method of Evaluating the Management Technique
5. Product Champion	Donaldson, Lufkin, & Janrette, Inc. (International Harvester Company) (3M Company) (Xerox Corporation) (International Telephone and Telegraph Corporation) (Donaldson, Lufkin & Janrette, Inc.) (Bell Laboratories) (The MITRE Corporation) (University of Wisconsin) (National Aeronautics and Space Administration) (University of Utah)	• Financial management and brokerage sources	• $1.5 million • 2,200 employees	• To identify someone who could be responsible for management and act as a "champion" for a new product or process, selling an idea to management and maintaining management's interest	• The new product or process becomes identified with the individual who personally reflects an interest in the development of the new product or process	• Measurement of profits associated with sales of new product or process
6. Project Performance Measurement	Technical Manufacturing Firm[b] (University of Utah)	• Glass and ceramic materials, components and systems for the chemical industry, automotive industry, scientific research, consumer products, laboratories, and optical, lighting, conservation, aviation and electronics industries	• $1.4 billion • 30,000 employees	• To establish guidelines to indicate whether an innovation project is being managed effectively	• Participation of individuals being evaluated in the development of the evaluation guidelines	• Measurement of relative decrease in management problems after implementation of the evaluation system
7. Quality Circles	Ford Motor Company (International Harvester Company)	• Automobiles, trucks, tractors, and implements	• $43.5 billion • 500,000 employees	• To improve quality of manufacturing to meet competitor's standards	• Use of outside consultants to train individuals responsible for	• Measurement of change in levels of productivity and creation of innova-

	Company	Financial	Industry	Purpose	Description	Measurement
(technical manufacturing firm)				and increase the flow of innovative ideas from the bottom of the organizational hierarchy upward	managing quality circles and preparation of an implementation schedule	tive ideas
8. Research Planning Frame	Xerox Corporation	• $8.2 billion • 120,000 employees	• Information processing, reprographics, and office systems	• To ensure consideration of corporate goals and philosophy in a long-range new product/service planning process	• Involvement of senior technical staff members on a rotating basis to mesh corporate and technical success	• Measurement of successful product development objectives
9. Temporary Groups	International Telephone and Telegraph Corporation (BF Goodrich Company) (3M Company) (Xerox Corporation) (Argonne National Laboratories)	• $18.5 billion • 368,000 employees	• Automotive and industrial products, food products and consumer appliances, processing timber and minerals, production of coal and oil, financial services and insurance	• To bring together individuals representing several levels of management to apply problem-solving skills to develop new product approaches	• The temporary nature of the group sets a time limit for producing results and the structure of the group permits cross-fertilization of ideas	• Measurement of satisfactory recommendations determined by their applicability and projected profitability
10. External Venture Management	Donaldson, Lufkin, and Jenrette, Inc. (University of Wisconsin) (National Aeronautics and Space Administration) (Argonne National Laboratories)	• $115 million • 2,200 employees	• Financial management and brokerage services	• To provide intensive capital, marketing and management resources to initiate new product development or assist mature firm's development and growth	• Services provided are expensive but highly effective, providing an excellent return on investment	• Measurement of return on investment
11. Franchising	McDonald's Corporation	• $6.2 billion	• Food service, franchise restaurants	• To establish a legal structure that will require participants in the system to maintain a standard level of quality and service	• Adhering to franchise guidelines ensures that franchise will benefit from technological innovations introduced through the corporate system	• Measurement of corporation's ability to meet the firms standards and increase technological adaptability to continuously improve the service

Table II, continued

Management Technique	Organization[a]	Organization's Products/Services	Organization's 1980 Revenues and Employment	Focus of the Management Technique	Factors Contributing to the Success of the Management Technique	Method of Evaluating the Management Technique
12. Licensing	Bell Laboratories (International Harvester Company) (Argonnne National Laboratories)	• Electronics and engineering research	• Not commercial • 20,000 employees	• To overcome the competition associated with the development of high technology by controlling communications among innovators in the field	• Active encouragement of information transfer regarding new product development	• Measurement of new product ideas received from licensees
13. Middleman Broker	The MITRE Corporation	• Research and engineering consultants	• $150 million • 3,000 employees	• To facilitate the matching of university services and resources with industry's technical innovation needs.	• The middleman, a not-for-profit organization, does not profit from any university-industry linkage and therefore acts as a surrogate for both participants	• Measurement of the number of technological innovations resulting from the match made by the middleman broker
14. Personnel Transfers	Bell Laboratories (National Aeronautics and Space Administration)	• Electronics and engineering research	• Not commercial • 20,000 employees	• Transfers were encouraged to create a feeling of goodwill among technical professionals among industry personnel	• Professional incentives were kept high, and communications with transferred personnel were encouraged to maintain a rapid pace in the development of the state-of-the-art technology	• Measurement of the rate of staff turnover
15. University-Industry Linkages	University of Wisconsin Small Business Development Center (National Aeronautics and Space Administration)	• Business management services	• Small core staff with consultants is drawn from the university	• To provide business management services to small businesses and individual inventors	• The business school at the university provides a wide range of consultant services by using the staff and students at the	• Measurement of types of contacts made and relative success resulting from center involvement as measured subjectively

				school as well as staff from the scientific and technical schools	by the person requesting the services
16. Industrial Applications Centers	• National Aeronautics and Space Administration	• Space research	• To assist in disseminating new technology developed for space research as widely as possible for the greatest public benefit	• The network established to facilitate the technology transfer enables public access to the vast library resources maintained by the administration	• Measurement of profits or cost savings resulting from use of a space technology product or process
17. Innovation Centers	• University of Utah Innovation Center	• Electrical equipment and electronics	• To provide startup funds for technical innovators to provide a physical location and management support services to entrepreneurs	• Small core staff, consultants are drawn from the university	• An individual with a strong personality and great experience in new venture management lent credibility to the program	• Measurement of the return on investment of projects sponsored by the center
18. National Technology Foundation	• National Technology Foundation	• Technical and engineering research	• To consolidate currently dispersed technical and engineering resources to address national problems	• Size and structure of this proposed federal-level agency have not been established	• Success or failure of the management technique undetermined	• Measurement of the success of this proposed management approach presumably will be conducted by measuring the ability of the organization to meet financial and program objectives (a standard method of evaluating federal-level agencies)
19. Regulations	• Health care industry	• Provision of health care services	• To stimulate the purchase and use of new health care technologies, regulations can be prepared to alter reim-	• Greater than $100 billion • The industry consists of myriad providers, most of	• The incentives must be structured to reward substantiated use, not overuse or overpurchase of a new technology	• Measurement of the value of a new technology ideally is determined by a change in the rate of morbidity

Table II, continued

Management Technique	Organization[a]	Organization's Products/Services	Organization's 1980 Revenues and Employment	Focus of the Management Technique	Factors Contributing to the Success of the Management Technique	Method of Evaluating the Management Technique
			whom are affected by government regulation	bursement or incentive rewards for users		and mortality experienced by the population served
20. Technology Transfer from Laboratories	Argonne National Laboratory (National Aeronautics and Space Administration)	• Research in the fields of nuclear energy, pressurized fluidized bed combustion and lithium-sulfide batteries	• Not commercial • 5000 employees	• To ensure effective commercialization of a technology that requires a large investment of R&D funds provided by the government	• Industry participants responded to the opportunity to learn about the new technology and provide continuous financial support for the training of individuals at the laboratory	• Measurement of the success of this management technique could be made by assessing the return on investment by the participating industry representative and shared R&D costs
21. Subsidies	Utility firms	• Nuclear power generation	• Greater than $100 billion	• To provide financial support to industries with high R&D costs, ensuring technological advancement and reduction of risk borne by technology developers	• A complete understanding of the secondary impacts of subsidies is required to ensure successful acceptance and operation of the subsidy	• Measurement of the success or failure of a subsidy and its primary and secondary impacts can not be conducted in terms of financial gain; the impact of a subsidy intended to stimulate innovation can be evaluated in terms of the return on investment of technologies developed as a result of the subsidy

[a]The first organization listed is the candidate organization, targeted in this study as representative of the subject management technique. The organizations listed in parentheses also have used or are using the subject management techniques.

[b]Because of the proprietary nature of the circumstances surrounding the application of this management technique, the firm cannot be identified.

Table III.   Phases During Which Management Techniques Are Employed

Management Technique	Phase I			Phase II			Phase III
	Goal Setting	Project Identification	Project Selection	Research	Development	Production	Diffusion
1. Idea Generation		↑					
2. Innovation Incentives				↑			
3. Innovation Training					↑		
4. Internal Venture Management							↑
5. Product Champion							↑
6. Project Performance Measurement						↑	
7. Quality Circles						↑	
8. Research Planning Frame			↑				
9. Temporary Groups			↑				
10. External Venture Management							↑
11. Franchising							↑
12. Licensing							↑
13. Middleman Broker							↑
14. Personnel Transfers							↑
15. University-Industry Linkages							↑

Table III, continued

Management Technique	Phase I			Phase II			Phase III
	Goal Setting	Project Identification	Project Selection	Research	Development	Production	Diffusion
16. Industrial Applications Center							
17. Innovation Centers							
18. National Technology Foundation							
19. Regulations							
20. Technology Transfer from Laboratories							
21. Subsidies							

Table IV.   Overview of the Groups of Barriers Affected by the Management Techniques

| | Barriers To Innovation | | |
Management Technique	Phase I Goal Setting Project Identification/Project Selection	Phase II Research/Development/Production	Phase III Diffusion
1. Idea Generation— Ford Motor Company	Technical, organizational, marketing	—	—
2. Innovation Incentives— International Harvester Company	Organizational	Technical, organizational	—
3. Innovation Training— B.F. Goodrich Company	Organizational	Organizational	—
4. Internal Venture Management— 3M Company	Organizational, marketing	Technical, organizational, financial, marketing	Organizational, financial, marketing
5. Product Champion— Donaldson, Lufkin & Jenrette, Inc.	Organizational, financial, marketing	Organizational, financial, marketing	Organizational, financial, marketing
6. Project Performance Measurement— a Technical Manufacturing Firm	Technical, organizational	Technical, organizational	—
7. Quality Circles Ford Motor Company	—	Organizational	—
8. Research Planning Frame— Xerox Corporation	Technical, organizational, financial, marketing	—	—
9. Temporary Groups— International Telephone and Telegraph Corporation	Technical, organizational, marketing	—	—
10. External Venture Management— Donaldson, Lufkin & Jenrette, Inc.	Organizational, financial	Organizational, financial	Financial, marketing
11. Franchising— McDonald's Corporation	Technical, financial	Technical, financial	Financial

Table IV, continued

Management Technique	Barriers To Innovation		
	Phase I Goal Setting Project Identification/Project Selection	Phase II Research/Development/Production	Phase III Diffusion
12. Licensing— Bell Laboratories	Technical	Technical, financial	Technical, governmental, financial
13. Middleman Broker— MITRE Corporation	—	—	Organizational, marketing
14. Personnel Transfers— Bell Laboratories	Technical, organizational	Technical, organizational, governmental	Technical, organizational, governmental
15. University-Industry Linkages— University of Wisconsin Small Business Development Center	Technical, financial	Technical, financial, marketing	Technical, organizational, financial, marketing
16. Industrial Applications Centers— National Aeronautics and Space Administration	Technical	Technical	—
17. Innovation Centers— University of Utah Innovation Center	Technical, organizational, financial	Technical, organizational, financial, marketing	Technical, organizational, financial, marketing
18. National Technology Foundation	Technical	Technical, organizational	Technical
19. Regulations	—	Governmental, financial	Governmental, financial
20. Technology Transfer from Laboratories	—	Technical, financial	Technical, financial
21. Subsidies	—	Financial	Governmental, financial

within the firm. The product champion and project performance measurement techniques have been used to overcome barriers associated with poor project organization and management. Such barriers inhibit innovation goal setting, as well as follow-through communications with top management.

Financial barriers that affect the first phase of the innovation process concern insufficient data on the financial risks associated with innovation and the inability to incorporate them in the corporate and project innovation program planning and budgeting activities. The product champion, the research planning frame and franchising address the problem of characterizing financial risks associated with innovation. The innovation centers program specifically addresses the problems associated with financial risk analysis in the development of business plans for new technologies. Other techniques identified in the study analysis, such as university-industry linkages, focus on overcoming the barriers associated with the lack of financial resources assigned to the innovation planning phase of the innovation process.

The management techniques that address marketing barriers within the first phase of the innovation process involve identifying persons responsible for establishing and coordinating marketing and technical goals early in the innovation planning process. The use of temporary groups is one approach to accomplish this aim. The groups also can assist in the early assessment of potential markets for the new technologies.

A total of 17 management techniques were identified as affecting barriers within the first phase of the innovation process. Tables V–VIII provide detailed summaries of the impact of these techniques on specific barriers. The relative impact of each management technique on the barriers has been ranked as "high" or "low," as defined by its users.

MANAGEMENT TECHNIQUES AFFECTING
BARRIERS WITHIN THE SECOND PHASE
OF THE INNOVATION PROCESS

The second phase consists primarily of the act of producing the new technology, or taking the idea out of the abstract and creating a salable product or process. Often, the individuals involved in this phase, such as design and production engineers, are different from those in the planning phase. The two phases do overlap, however, because many technological advances are discovered during the R&D, or second stage, of the innovation process. The barriers found to exist during this phase of the process

Table V. Management Techniques Affecting Technical Barriers During Phase I of the Innovation Process[a]

Management Technique	Technical Barriers		
	Lack of Technical Information	Lack of Support for Collateral Invention	Lack of Other Technical Resources
1. Idea Generation	● "Low" brings together technical manpower resources to address new product/process formulation problems.	—	—
2. Project Performance Measurement	—	● "High" encourages ongoing evaluation of new product ideas involving innovators and innovation managers	—
8. Research Planning Frame	● "High" requires the involvement of technical entrepreneurs and managers in the product/process planning process	—	—
9. Temporary Groups	● "High" encourages the participation of technical managers from several levels of the corporate hierarchy to participate in the new product planning process.	—	—
11. Franchsing	—	—	● "High" maintains all technical resources required to support innovation at the central corporate level.

12. Licensing	• "High" provides a mechanism for the transfer of new product/process ideas to firms responsible for research and development.	• "High" encourages the transfer of new product/process ideas among licensee network participants	• "High" centralizes technical resources required to stimulate the planning of innovation.
14. Personnel Transfer	• "High" facilitates the rapid transfer of innovative ideas among industry participants encouraging the development of new products and processes.	• "High" stimulates the meshing of technical talents required to maintain the momentum of innovation.	• "High" assures the maintenance of a competitive atmosphere within industry to provide technical innovators with required technical resources.
15. University-Industry Linkages	• "High" matches the identifiers of market needs (industry) with researchers (university) to facilitate setting priorities among new product/process development plans.	—	—
16. Industrial Applications Centers	• "High" disseminates new product and process concepts to industry.	• "High" matches technological resources with industrial needs to define new applications for products and processes.	• "High" provides previously developed technology to affect newly defined market needs.
17. Innovation Centers	• "High" provides professional technical assistance to entrepreneurs for the development of innovative ideas.	• "High" creates a low-risk supportive environment for innovators to evaluate new product/process plans.	• "High" provides technical resources required to stimulate the movement of entrepreneurs from thought to action.
18. National Technology Foundation	• "High" creates an organizational structure supportive of informa-	• "High" provides a setting for professionals to interact to define	• "High" pools the technical resources that traditionally have

Table V, continued

Management Technique	Technical Barriers		
	Lack of Technical Information	Lack of Support for Collateral Invention	Lack of Other Technical Resources
	tion transfer among professions, which is required to stimulate technological innovation.	new technological innovation	been segregated to increase access by professionals.
19. Technology Transfer from Laboratories	• "High" instills a strong bond between innovators (laboratories) and R&D/marketing participants (industry) to improve focus on purpose of new technology.	• "High" provides a setting to enable laboratory researchers to work with industry representatives to develop and evaluate new product/process concepts.	• "High" provides technical resources along with technical consultation to industry representatives for the purpose of assessing new product/process ideas.

[a]The outlined sections indicate the primary focus of the management technique. "High" and "low" rankings have been assigned to the individual activities associated with the techniques to indicate their relative impact on given barriers.

Table VI. Management Techniques Affecting Organizational Barriers During Phase I of the Innovation Process[a]

Management Techniques	Lack of Support From Top Management	Environment Not Supportive of Innovators	Organizational Barriers	Lack of Adequate Professional Recognition and Reward	Ineffective Project Management
			Inefficient Communication Channels		
1. Idea Generation	• "High" involves top management in the development and evaluation of the product/process ideas.	• "Low" encourages part-time use of innovators in idea-generating sessions.	• "High" involves innovators in idea generating sessions that bring about synergistic response to the identification of new products and processes.	—	—
2. Innovation Incentives	• "High" involves upper management in establishing rewards for the discovery of new product/process ideas.	—	• "Low" stimulates the rate of innovation and supporting communications infrastructure as a result of incentives awards.	• "High" encourages the development of new technological innovation through a reward structure.	—
3. Innovation Training	• "High" requires that top management approve and participate in the curriculum development.	• "High" provides training for innovators and encourages participation in the innovation process.	• "High" creates communications linkages designed specifically to facilitate innovation.	—	—
4. Internal Venture Management	• "High" requires identification of key personnel to manage innovation projects during early stages of the process.	• "High" establishes an isolated environment in which innovators can carefully evaluate new product concepts.	—	• "Low" permits assignment of risks (and later rewards) to individuals involved early in the process.	—
5. Product Champion	• "High" identifies individual to present new product concept to top management.	—	—	—	• "High" ensures that one individual will have the responsibility of following the new product concept through all phases of the innovation process.
6. Project Performance Measurement	—	—	—	—	• "High" enables management to evaluate innovation project selection process to

Table VI, continued

Techniques	Organizational Barriers				
	Lack of Support From Top Management	Environment Not Supportive of Innovators	Inefficient Communication Channels	Lack of Adequate Professional Recognition and Reward	Ineffective Project Management
8. Research Planning Frame	• "High" involves top management in long-range innovation project planning.	—	• "High" involves technical staff and management representatives in the development of corporate goals.	—	identify and correct management problems.
9. Temporary Groups	• "High" includes representatives of top management in goal setting activities conducted by temporary groups.	—	• "High" establishes new communications channels by bringing together individuals from different management levels and areas of technical expertise.	—	—
10. External Venture Management	• "High" provides top management to poorly managed businesses developing new technology.	—	• "High" provides management support to encourage communications supportive of innovation.	—	• "High" utilizes highly qualified management experts/consultants to ensure effective project management.
14. Personnel Transfers	—	—	—	—	• "High" enables recruitment of highly qualified technical managers from within or outside the organization.
17. Innovation Centers	—	• "High" establishes a protective environment enabling entrepreneurs to seek technical assistance regarding project evaluation.	—	—	• "High" provides management services to assist the entrepreneur in development of a new product business plan.

aThe outlined sections indicate the primary focus of the management technique. "High" and "low" rankings have been assigned to the individual activities associated with the techniques to indicate their relative impact on given barriers.

Table VII. Management Techniques Affecting Financial Barriers During Phase I of the Innovation Process[a]

Management Technique	Financial Barriers		
	High Financial Risk Associated with the Process Phase	Inability to Use Forecasts and Estimates	Lack of Financial Resources Assigned to Innovation
5. Product Champion	• "High" individual representing new product concept attempts to quantify risk associated with product development.	—	• "Low" ensures continuous project funds monitoring—by product champion—to identify financial problems before project activities are affected.
8. Research Planning Frame	• "High" involvement of top management in planning of innovation program ensures consideration of ranges of risks to be assumed by the firm with regard to the innovation process.	• "High" involvement of top management in a structural innovation planning process ensures use of planning forecasts and estimates.	• "Low" involvement of top management in planning of innovation program increases probability of adequate funds being assigned to innovation projects.
11. Franchising	• "High" assumption of innovation risks by franchisor lessens the risks for all participants in the franchise network.	—	• "High" pooling of financial resources at the corporate level enables more focused R&D activities to be considered and undertaken.
15. University-Industry Linkages	—	—	• "Low" permits concentration of research funds to apply to specific technical innovation projects.

Table VII, continued

Management Technique	Financial Barriers		
	High Financial Risk Associated with the Process Phase	Inability to Use Forecasts and Estimates	Lack of Financial Resources Assigned to Innovation
17. Innovation Centers	—	• "Low" ensures proper consideration of forecasts and estimates in product planning by involving business school students and staff to assist in the preparation of business plans.	—

[a] The outlined sections indicate the primary focus of the management technique. "High" and "low" rankings have been assigned to the individual activities associated with the techniques to indicate their relative impact on given barriers.

Table VIII. Management Techniques Affecting Marketing Barriers During Phase I of the Innovation Process[a]

Management Technique	Marketing Barriers	
	Lack of Information Concerning Characteristics of Potential Markets	Inability to Coordinate Technical and Marketing Goals
1. Idea Generation	• "High" involves technical experts who are cognizant of market characteristics in the product planning phase.	—
4. Internal Venture Management	—	• "Low" establishes a special group of individuals, including technical and marketing representatives, to participate in new product planning.
5. Product Champion	—	• "High" requires that one individual be responsible for overseeing the integration of technical and marketing goals during new product planning.
8. Research Planning Frame	• "Low" involves technical and marketing experts in the creation of new product market plans.	• "Low" involves technical and marketing experts in the formal planning of new products and processes.
9. Temporary Groups	• "Low" temporarily draws market experts into the new product planning process.	• "Low" encourages the interaction of technical and marketing management representatives to establish agreed on new product goals.

[a]The outlined sections indicate the primary focus of the management technique. "High" and "low" rankings have been assigned to the individual activities associated with the techniques to indicate their relative impact on given barriers.

fell into all five of the groups—technical, organizational, governmental, financial and marketing barriers.

Technical barriers impacting on research, development and production of new technology occur for one of two reasons: (1) an absence of technical qualifications within the firm to undertake the activities; or (2) an inability to foster innovation with existing technical resources. Licensing, personnel transfers and industrial applications centers address these problems by providing technical expertise and stimulation to the firm. Outside innovation centers, the National Technology Foundation and technology transfer from a national laboratory are management techniques that have been utilized to create technical support for the research, development and production of new technology.

The organizational barriers affect this phase by impeding the development of the new technology in failing to organize resources properly. Innovation training, internal venture management and quality circles attempt to improve communications regarding the R&D of innovation to increase the awareness of the need for top management to continually support the development of the new technology. The product champion and innovation centers are examples of management techniques that focus the attention of the corporation or investors on innovation developments.

Governmental barriers may impede this phase by causing management to doubt the innovation's eventual worth. Such doubts can be instilled through the threat of burdensome regulatory policies or the implication of a lack of patent protection. Government policy may be used not only to inform and assist innovators in the development of new technology, but to actually provide financial support for specific R&D.

Financial barriers in the second phase are related to monetary risks associated with research, development and production. This phase is often the most costly and frequently calls for high-risk capital. The management techniques that address these financial risk-related problems (internal venture management, product champion, franchising, and licensing) require that the financial risks associated with the other phases be assessed and monitored. Thus, decision-makers would be made aware of the costs associated with the development of the new technology and be able to evaluate the risk associated with the innovation project.

Marketing barriers are relatively minor during the second phase of the innovation process. Nonetheless, management techniques such as internal venture management, product champion, industrial applications centers and innovations centers have focused on the need to maintain an ongoing check of the match between technical and marketing goals as the innovation project evolves.

Seventeen management techniques were identified in the study analysis as having impacts on barriers that occur during the research, development and production phases of the innovation process. Tables IX–XIII present these management techniques as they affect specific barriers. The relative impact of each management technique on the specific barriers included in the analysis has been ranked "high" or "low" as defined by the users of the technique or as evaluated by the interviewer.

MANAGEMENT TECHNIQUES AFFECTING BARRIERS WITHIN THE THIRD PHASE OF THE INNOVATION PROCESS

The third and final phase of the innovation process involves the many operations necessary to achieve diffusion of the new technology. The management techniques affecting barriers within this phase of the process concentrate on the financial and marketing barriers, although technical, organizational and governmental barriers also affect the diffusion process.

The technical barriers affecting diffusion result from the lack of information on the new technology, which, in turn, may result from the division of responsibility between phases one and two without adequate technical interfaces. This problem was notably affected by the personnel transfer and technology transfer from laboratory techniques, which facilitate the liaison between research/development/production and diffusion.

Organizational barriers, as in the previous phases of the process, can be characterized as blocked relationships, or communications channels that inhibit technological innovation. During the diffusion phase, the management techniques have their major impacts on the organizational barriers associated with ineffective project management. The middleman/broker, the university-industry linkages and other techniques establish formal information and organization relationships to support technological innovation.

Governmental barriers impacting technological diffusion were addressed by licensing, personnel transfers, the establishment of regulations and government subsidies. These techniques have assisted in the diffusion process by providing financial and technical assistance to overcome uncertainties regarding government policy and activities.

Several management techniques address financial and marketing barriers within the diffusion phase of the innovation process. Because there is often a sizable financial commitment to market a new product, risks

Table IX.   Management Techniques Affecting Technical Barriers During Phase II of the Innovation Process[a]

Management Technique	Technical Barriers		
	Lack of Technical Information	Lack of Support for Collateral Invention	Lack of Other Technical Resources
1. Innovation Incentives	—	• "Low" creates specific innovation project goals for inventors and encourages collateral innovation.	—
4. Internal Venture Management	—	• "Low" stimulates collateral invention by isolating innovators from pressures or distractions from the firm's normal operations.	—
6. Project Performance Measurement	—	• "Low" establishes, through group goal setting, an environment conducive to collateral invention.	—
11. Franchising	—	—	• "High" provides all levels of technical resources at the corporte level to be applied to R&D projects.
12. Licensing	• "High" ensures provision of technical information as part of the licensing agreement.	• "High" encourages collateral invention through technical information exchange between licensing agent and licensees.	• "High" ensures provision of technical resources as part of the licensing agreement.

14. Personnel Transfers	• "High" enables transfer of technical information regarding R&D projects.	• "High" encourages individuals to pursue opportunities involving collateral invention.	• "High" ensures provision of technical resources in support of personnel transfers.
15. University-Industry Linkages	• "Low" establishes communications linkages that permit the flow of technical information between R&D groups.	—	—
16. Industrial Applications Centers	• "High" draws on a highly technical information base to address R&D problems in industry.	• "High" provides technical assistance to stimulate collateral inventions.	• "HIgh" identifies other sources of technical resources within the government and industry to assist in the new product R&D.
17. Innovation Centers	• "High" provides the entrepreneur access to technical and management resources at the University.	• "High" establishes a physical and technical environment supportive of collateral invention.	• "High" provides the entrepreneur with access to technical resources within the University and business communities.
18. National Technology Foundation	• "High" establishes a formal funding and technical management structure to relay technical information for new product R&D.	• "High" provides a setting for technical experts to converge to address common problems.	• "High" establishes a central focus for inventorying and providing access to technical resources required for technological innovation.
20. Technology Transfer from Laboratories	• "High" involves industry in the R&D of highly technical projects.	• "High" encourages participation of industry representatives in the problem-solving aspect of new product development.	• "High" concentrates on the technical resources required for R&D on one project area.

[a] The outlined sections indicate the primary focus of the management technique. "High" and "low" rankings have been assigned to the individual activities associated with the techniques to indicate their relative impact on given barriers.

Table X.   Management Techniques Affecting Organizational Barriers During Phase II of the Innovation Process[a]

Management Technique	Organizational Barriers				
	Lack of Support from Top Management	Environment Not Supportive of Innovators	Insufficient Communications Channels	Lack of Adequate Professional Recognition and Reward	Ineffective Project Management
1. Innovation Incentives	—	—	—	• "High" establishes a reward structure and career ladders for successful R&D activities, as well as management approaches.	—
3. Innovation Training	• "High" requires top management to participate in an innovation training program involving them in the management of research, development and production.	• "High" uses corporate managers trained in organizational development techniques to provide a supportive and challenging environment for innovators.	• "High" increases communication linkages supporting innovation in R&D.	—	—
4. Internal Venture Management	• "High" involves top management in all stages of new venture management.	• "High" sets priorities for the establishment of a climate conducive to innovative R&D.	—	• "Low" assigns risks (and later rewards) to individuals involved in new venture management during research, development and production of new products/processes.	—
5. Product Champion	—	• "High" individual acts as advocate for new project, safeguarding the creative environment for the innovation.	—	—	• "High" ensures active interest on the part of project, as well as top management, in project progress.
6. Project Performance Measurement	—	—	• "High" establishes formal communications lines among management and technical	—	• "High" enables management to evaluate relative success of project operation based on operational variables.
7. Quality Circles	• "Low" involves top management in the evaluation of recommendations pertaining	—		—	—

Technique			
	to research, development, and production improvements.	*staff and encourages the flow of creative ideas.*	
10. External Venture Management	"High" creates a staff of top management specifically supportive of innovation activities.	—	—	"High" highly prioritizes the assignment of project management with the technical and organizational skills to enhance the project's growth and development.
14. Personnel Transfers	—	"High" encourages the development of strong and effective communication channels to ensure achievement of innovation project objectives.	"Low" encourages rapid exchange of technical information through word of mouth versus written communications.	"High" permits introduction of qualified technical management to be drawn from other parts of the firm or from outside the firm.
15. University-Industry Linkages	—	—	"High" enables individuals to seek rewards or recognition to satisfy professional needs from alternative sources.	"High" enables project management to be drawn from multiple sources—university or industry based.
17. Innovation Centers	—	"High" establishes an environment solely to support the activities of the entrepreneur.	—	"High" provides management services to assist the entrepreneur in the development of the new product or process from its inception through its diffusion in the market.
18. National Technology Foundation	—	—	—	"High" identifies key financial leaders to serve in the role of managing diverse multidisciplinary resources to solve national social and technical problems.

[a] The outlined sections indicate the primary focus of the management technique. "High" and "low" rankings have been assigned to the individual activities associated with the techniques to indicate their relative impact on given barriers.

Table XI. Management Techniques Affecting Governmental Barriers During Phase II of the Innovation Process[a]

Management Technique	Governmental Barriers	
	Uncertainty Concerning Regulatory Policies	Inadequate Patent Protection
14. Personnel Transfers	—	• "Low" enables individuals to protect potentially threatened patentable ideas by leaving the firm.
19. Regulations	• "Low" increases understanding concerning impacts of alternative incentives programs.	—

Table XII.   Management Techniques Affecting Financial Barriers During Phase II of the Innovation Process[a]

Management Technique	High Financial Risk Associated with Process Phase	Financial Barriers	
		Inability to Use Forecasts and Estimates	Lack of Financial Resources Assigned to Innovation
4. Internal Venture Management	• "Low" obviates high financial risk in R&D as early project planning has permitted selection of acceptable risk projects.	—	• "Low" establishes financial investment in the R&D activities early in the project planning phase.
5. Product Champion	• "High" educates top management (decision-makers) regarding the nature of risks associated with the research and development phase and presents alternative funding options.	—	• "Low" maintains interest of top management in financially supporting the innovation project.
10. External Venture Management	• "High" assigns risk associated with investment in projects early in the innovation process so the realization of risk at this second phase of the process does not act as a deterrent to the success of the project.	—	• "High" assigns appropriate financial resources as needed based on early commitment to project and return on investment requirements.
11. Franchising	• "High" provides for innovation risks to be borne by corporate headquarters, rather than by small businessmen.	—	• "High" establishes an R&D budget to support ongoing technological innovation.

Table XII, continued

Management Technique	Financial Barriers		
	High Financial Risk Associated with Process Phase	Inability to Use Forecasts and Estimates	Lack of Financial Resources Assigned to Innovation
12. Licensing	• "High" enables risk to be assumed by an organization more in a position financially and technically to assume risks associated with technological research and development than the licensees, which are generally smaller firms.	—	• "High" encourages the concentration of financial resources to permit support of R&D.
15. University-Industry Linkages	—	—	• "Low" encourages the concentration of financial resources to permit support of R&D within the university setting.
17. Innovation Centers	• "High" establishes budget to support R&D based on evaluation of innovation project conducted earlier in the innovation process.	—	• "High" ensures commitment of investment funds throughout all phases of the innovation process.
19. Regulations	• "High" government, a major investor in innovation, seeks to overcome financial risks associated with R&D, rather	—	• "High" establishes a relatively long-term horizon for providing financial support for R&D in the private sector.

than the individuals actually performing it.

| 20. Technology Transfer from Laboratories | ●"High" government labs assume financial risk for R&D of innovations to be commercialized publicly. | ●"High" financial resources are designated for specific technological innovation projects as part of larger R&D efforts. |
| 21. Subsidies | ●"High" enables financial risks associated with R&D to be borne by government. | ●"High" establishes a pool of financial resources earmarked to support the R&D of certain technological innovation projects. |

[a]The outlined sections indicate the primary focus of the management technique. "High" and "low" rankings have been assigned to the individual activities associated with the techniques to indicate their relative impact on given barriers.

Table XIII. Management Techniques Affecting Marketing Barriers During Phase II of the Innovation Process[a]

| | Marketing Barriers | |
Management Technique	Lack of Information Concerning Characteristics of Potential Markets	Inability to Coordinate Technical and Marketing Goals
4. Internal Venture Management	—	• "Low" ensures consideration of coordinated technical and marketing goals through consistent project management following new product development.
5. Product Champion	—	• "High" ensures consistent coordination of technical and marketing goals through role of product champion.
16. Industrial Applications Centers	• "Low" utilizes access to data bases to assess potential markets.	• "Low" works closely with industry and small businesses to identify technical and marketing goals.
17. Innovation Centers	—	• "Low" provides management support to entrepreneurs throughout the innovation process to ensure coordination of technical and marketing goals.

associated with the investments in the diffusion activities must be measurable and manageable. Management techniques such as internal venture management, external venture management, the middleman broker and university-industry linkages seek to minimize risks by assessing market needs and matching them with new technological products and processes.

Of the 21 management techniques surveyed, 14 addressed barriers within the technological diffusion phase. Tables XIV–XVIII indicate the barriers affected by these management techniques. The relative impact of each management technique on given barriers included in the analysis has been ranked "high" or "low" as defined by the users of the technique or as evaluated by the interviewer.

SUMMARY

This chapter describes a sample of management techniques that have proved to be effective tools in facilitating the technological innovation process. Although these techniques have had a positive influence on the organizations described, they will not necessarily produce the same results in other industrial settings. However, key elements or characteristics of the management techniques that have led to their success are identified here. Some of these elements or characteristics supporting successful innovation are:

- the existence of, and accessibility to, state-of-the-art technology and technologists;
- the presence of an individual who is interested in seeing the technological innovation nurtured through the process;
- the continuous support and ongoing development of an organizational structure conducive to innovation providing the required technical, financial and management resources;
- the development of government policies and regulations that stimulate (or, at a minimum, do not deter) innovation.
- The ability to control, budget and evaluate the process in line with short- and long-term organizational goals.

The views of a sample of management techniques currently used in U.S. industry and government to facilitate technological innovation also are summarized. Continued study of specific applications of these management techniques would reveal useful information regarding transfer of the techniques to other industrial settings.

Table XIV.   Management Techniques Affecting Technical Barriers During Phase III of the Innovation Process[a]

Management Technique	Technical Barriers		
	Lack of Technical Information	Lack of Support for Collateral Invention	Lack of Other Technical Resources
12. Licensing	—	—	• "High" provides technical resources as part of the licensing agreements to ensure effective product diffusion.
14. Personnel Transfers	• "High" increases the speed by which technical information (new technology) is diffused.	—	• "High" provides technical support and services required to market new products and processes.
15. University-Industry Linkages	• "Low" establishes a technical information base required to facilitate technological diffusion.	—	—
17. Innovation Centers	• "High" provides technical information from engineering and business schools to support technological diffusion.	—	• "High" provides other technical resources from industry and community sources to ease new product diffusion.
18. National Technology Foundation	• "High" establishes a central focus for technical information in support of the diffusion of new technology.	—	• "High" encourages the identification and application of other technical resources to support the diffusion of new technology.

20. Technology Transfer
    from Laboratories

- "High" involves industry participants responsible for the diffusion of technology in the early stages of R&D to increase understanding of the technology.

—

- "High" utilizes technical resources of government facilities and industry to ensure product diffusion.

[a] The outlined sections indicate the primary focus of the management technique. "High" and "low" rankings have been assigned to the individual activities associated with the techniques to indicate their relative impact on given barriers.

Table XV. Management Techniques Affecting Organizational Barriers During Phase III of the Innovation Process

Management Technique	Organizational Barriers				
	Lack of Support from Top Management	Environment Not Supportive of Innovators	Insufficient Communications Channels	Lack of Adequate Professional Recognition and Reward	Ineffective Project Management
4. Internal Venture Management	• "High" requires that top management assign responsibility and financial resources to new product marketing.	• "High" establishes a group within the organization to nurture the innovation through the diffusion phase of the innovation process.	—	• "High" provides financial rewards for management of new product groups.	—
5. Product Champion	• "High" ensures innovation support by having an advocate of the new concept participate in corporate market planning.	—	—	—	• "High" use of a product advocate or champion ensures that one person will remain active and be responsible for the diffusion of the new product.
9. Temporary Groups	• "High" involves top management in the product/process diffusion phase.	—	• "High" strengthens communications linkages by involving several levels of management in the group.	—	—
13. Middleman/ Broker	—	—	—	—	• "High" uses a coordinating institution to manage project diffusion, obviating political influences of participating organizations.
14. Personnel Transfers	—	—	• "Low" increases creation of communications linkages through the establishment of new relationships among professionals resulting from the physical transfer of personnel.	• "High" encourages individuals to gain recognition and reward for assisting in technological diffusion by seeking new positions.	—

15. University-Industry Linkages	—	—	—	• "High" establishes a formal mechanism or management responsibility for the diffusion of new products/processes.
17. Innovation Centers	—	• "High" establishes a productive environment enabling entrepreneurs to seek technical assistance in the diffusion of new products and processes.	—	• "High" provides professional management assistance for all phases of the innovation process.

[a]The outlined sections indicate the primary focus of the management technique. "High" and "low" rankings have been assigned to the individual activities associated with the techniques to indicate their relative impact on given barriers.

Table XVI. Management Techniques Affecting Governmental Barriers During Phase III of the Innovation Process[a]

	Governmental Barriers	
Management Technique	**Uncertainty Concerning Regulatory Policies**	**Inadequate Patent Policy**
12. Licensing	•"Low" increases information flow among licensee network regarding anticipated regulatory policies.	•"High" encourages sharing of technical information and establishes respect for patents through licensing.
14. Personnel Transfers	—	•"Low" enables innovators to guard patents by taking new product ideas to another firm for development and diffusion.
19. Regulations	•"Low" establishes framework for anticipated government policies.	—
21. Subsidies	•"High" establishes guidelines for subsidizing the marketing or diffusion of new technologies. Supportive regulating policies may be implied.	—

[a]The outlined sections indicate the primary focus of the management technique. "High" and "low" rankings have been assigned to the individual activities associated with the techniques to indicate their relative impact on given barriers.

Table XVII. Management Techniques Affecting Financial Barriers During Phase III of the Innovation Process

Management Technique	Financial Barriers		
	High Financial Risk Associated With Process Phase	Inability to Use Forcasts and Estimates	Lack of Financial Resources Assigned to Innovation
4. Internal Venture Management	—	—	• "Low" establishes financial investment supportive of product diffusion.
5. Product Champion	• "High" educates top management regarding the nature of risks associated with diffusion of the product.	—	• "Low" continues to seek required technical resources facilitating product diffusion.
10. External Venture Management	• "High" assigns risk associated with investment in the project early in the innovation process so the realization of risk during the product diffusion stage does not serve as a deterrent to project success.	—	• "High" assigns appropriate financial resources as needed for product diffusion based upon early commitment to project and return on investment requirements.
11. Franchsing	• "High" provides for financial risks associated with product diffusion to be borne by the franchise corporate headquarters rather than the small businessmen franchisees.	—	• "High" establishes a financial support base to ensure diffusion of new technologies developed by the franchise corporate headquarters.

Table XVII, continued

| | | Financial Barriers | |
Management Technique	High Financial Risk Associated With Process Phase	Inability to Use Forcasts and Estimates	Lack of Financial Resources Assigned to Innovation
12. Licensing	• "High" risks associated with diffusion are lessened by virtue of technical support provided through the licensing agreement.	—	• "High" ensures reservation of funds as licensees have not had to spend large amounts to support R&D projects.
15. University-Industry Linkages	—	—	• "Low" enables industry to focus on financial support for diffusion of all technology rather than the R&D effort.
17. Innovative Centers	• "High" establishes budget to cover financial risks identified during the evaluation of project feasibility conducted during the first phase of the innovation process.	—	• "High" ensures commitment of investment funds throughout all phases of the innovation process.
19. Regulations	• "High" assumes financial risks associated with diffusion of technology by providing reimbursement for use.	—	• "High" establishes financial base to pay for reimbursement for services rendered.

20. Technology Transfer from Laboratories	• "High" Financial risks associated with diffusion of technology borne by industry are decreased as a result of market analysis efforts conducted concurrently with R&D.	—	• "High" ensures financial resource commitment for product diffusion by industries participating in the technology development project.
21. Subsidies	• "High" enables financial risks associated with the diffusion of technology to be borne by government.	—	• "High" establishes a pool of financial resources to be used for the diffusion of certain technological innovation projects.

[a]The outlined sections indicate the primary focus of the management technique. "High" and "low" rankings have been assigned to the individual activities associated with the techniques to indicate their relative impact on given barriers.

Table XVIII. Management Techniques Affecting Marketing Barriers During Phase III of the Innovation Process[a]

Management Technique	Marketing Barriers	
	Lack of Information Concerning Characteristics of Potential Markets	Inability to Coordinate Technical and Marketing Goals
4. Internal Venture Management	—	• "Low" establishes a high priority on the evaluation of technical and marketing goals early in the project operation to ensure continuity.
5. Product Champion	—	• "High" maintains coordination of technical and marketing goals through the product champion.
10. External Venture Management	—	• "High" focuses a high level of management and technical staff effort on the diffusion of new products.
13. Middleman/Broker	• "High" serves to assess the characteristics of the potential markets to be addressed by the new technology.	• "High" acts as coordinator for university and industry providing technical and marketing resources.
15. University-Industry Linkages	• "Low" draws resources from the university and industry communities to assess potential markets for the new technology.	• "Low" establishes mechanisms to maintain open communications regarding technical and marketing goal-setting.
17. Innovation Centers	—	• "High" conducts marketing research to accompany development of new products and processes.

[a]The outlined sections indicate the primary focus of the management technique. "High" and "low" rankings have been assigned to the individual activities associated with the techniques to indicate their relative impact on given barriers.

CHAPTER 4

DESCRIPTIONS OF MANAGEMENT
TECHNIQUES

IDEA GENERATION—
FORD MOTOR COMPANY

Introduction

Background

Large and small corporations must respond to the requirements of technological innovation within their respective industries. To produce a continuous stream of new products or process ideas, many formal and informal methods of generating ideas are employed. One broad group of idea generation techniques may be referred to as "brainstorming," which may be a structured or relatively unstructured mechanism of creating a communications flow, usually among technological innovators and corporate financial decision-makers. This management technique is used to overcome barriers associated with the lack of communication among the innovators within the traditional corporation hierarchy.

Organization

The Ford Motor Company was one of the original users of the idea-generating method called the Delphi technique. This technique draws on the expertise of individuals with considerable experience in a given industry. It requires them to make quantitative and qualitative projections of future new products and market needs. This technique has been particularly attractive in the automotive industry as it calls on many different organizational levels and areas of expertise to develop estimates or projections regarding analytically unanswerable questions. The operation of the Delphi technique, as well as other idea-generating methods, is discussed below. These approaches to identify new products have been used in organizations of all sizes to stimulate creativity and decision-making. The organizational key to applying idea generating techniques is flexibility.

Management Technique

Overview

Brainstorming, a precursor of the Delphi technique, was formally introduced as a management technique in 1938 and gained great acceptance in the 1950s, particularly in highly creative areas such as advertising and marketing. The term brainstorming refers to gathering creative individuals in a free form meeting, without critical review, to seek solutions to problems. The expected result of a brainstorming session is that more creative ideas will emerge through the synergistic reaction of the group than if the individuals expended equivalent effort without benefit of the group's interaction.

The general guidelines that assist in the successful conduct of this management technique are as follows:

- The group of creative innovators is given a single problem on which to focus. To avoid dissipation of the group's energy, multiple related problems are not presented.
- One of the group's goals is to encourage the suggestion of numerous ideas regarding possible solutions to the problem.
- Evaluation of the individual ideas should be delayed until all suggested ideas regarding the solution to the problem have been presented.
- Development of the proposed solution(s) should not be conducted by this group.
- Individuals who might inhibit the development of creative ideas, possibly key managers or supervisors, should be excluded from the brainstorming session.
- Brainstorming sessions should be set in a relaxed atmosphere rather than a work setting.

Selection and Implementation of the Management Technique

Brainstorming groups should be kept small to permit an easy exchange of ideas. The group also should consist of representatives of different disciplines important to the innovation process, such as engineers, technicians, marketing professionals and, possibly, sales personnel. The group usually is directed by a chairperson who outlines the goal or purpose of the brainstorming session and maintains the focus of the discussion on the problem.

Variations of the technique are called buzz groups, the Delphi technique and synectics. Buzz groups, which also may be referred to as free association groups, multiple brainstorming, reverse brainstorming or secret brainstorming were developed by Donald Phillips to increase participation in problem-solving sessions. Buzz groups are created by having a large group of individuals, similar in makeup to the brainstorming group, count off by sixes to form small groups. These groups elect a chairman and secretary and meet for short periods (e.g., six minutes) to identify a potential solution for a given problem. At the end of this session, each chairperson announces the ideas proposed by his or her group to the larger group. The larger group then evaluates the proposed solutions. The primary objective of the buzz group is to overcome the organizational problems associated with other methods of group formation: aggressive individuals overtake the meeting, not allowing others to participate. By forming buzz groups, the aggressive individuals will be randomly placed in small groups.

The Delphi technique was developed by the Rand Corporation, a research and engineering firm, to solicit answers to specific questions or problems. It differs from the previously described technique in that professionals are approached individually and asked to suggest alternative solutions. The ideas are collected centrally, summarized and statistically analyzed in terms of frequency and category of response. The analysis of ideas or list of potential solutions is then distributed to the individuals for their review. Based on the consolidated summary of responses, the individual is asked to revise his or her answer, if necessary. The review process may be repeated up to three times to permit final selection of a feasible alternative.

Synectics is a management technique developed by a consultant training firm to generate new product or process ideas by using team development skills and by creating a stimulating environment. This technique is similar to brainstorming as it includes a small group of five to seven individuals who represent diverse technical backgrounds. Like brainstorming, the synectics method encourages the deferral of a critical examination of the suggested ideas until the creative session has drawn to a close. Synectics differs from brainstorming, however, in placing a greater emphasis on the support of the individuals and their roles in the group dynamics process. The group participants actually are trained to develop problem-solving skills, which may be used in a number of different settings.

Evaluation of the Management Technique

The major advantage of using idea generation in a group setting is that the method usually encourages a number of new product or process ideas to be created rapidly. The group generally produces more ideas than the individuals could working by themselves. The major problems associated with the application of idea-generating management techniques, or the reason why they fail, is that the groups are not formed carefully and the guidelines that accompany the group's development are not followed. This problem is usually caused by the lack of interest in, or knowledge of, the technique. Another problem is the inability of top management to define the problem clearly, or of the group leader to convey the full meaning of the problem and specify a desired range of solutions. Another difficulty is that if the organizational ranks of the group participants vary too greatly, creativity is hindered by the unsolved power structure.

The group concepts described here generally attempt to create a climate appropriate for idea generation. Factors supportive of a creative environment suggest that management should:

- recognize that innovators require operational settings without limiting time schedules;
- permit innovators to associate freely with others in a similar position;
- recognize achievements of innovators; and
- establish challenges of innovators.

Because idea-generators or innovators require stimulation of creative responses so that they continue to be challenged, management should establish an environment and idea-generating structure within the organization that will permit the continuous identification of new challenges.

Other organizations and institutions using this technique are as follows:

- Bell Laboratories
- The MITRE Corporation
- University of Utah
- National Technology Foundation

INNOVATION INCENTIVES—
INTERNATIONAL HARVESTER COMPANY

Introduction

Background

Within the industrial research and development setting, technical professionals often must assume management responsibilities to advance and gain greater recognition and financial rewards. Unfortunately, a technical professional may have neither the skills nor the desire to enter management. As a result of promotion, the company often acquires an ineffective manager and loses the skills and innovative potential from the technical professional. Some industrial firms have addressed this problem by establishing dual career ladders, which permit, and in some cases encourage, an individual to pursue a career in either management or a technical area. The company provides comparable financial rewards for both career ladders.

The compensation program described here was developed at the International Harvester Company. It differs from other corporate dual ladder systems because in addition to parallel salary increases, it provides special financial rewards for individual professional achievement comparable to those offered to supervisory/management personnel, as well as an opportunity for individuals to gain special recognition within the firm.

The management technique described in this case study was developed and implemented during the past three years by a senior officer of a large industrial corporation. The technique is a compensation program aimed at attracting and retaining top technical professionals. The program includes incentives for technical staff members at different stages of their careers. It encourages the continued development of technical skills, while providing the impetus for innovators to develop new products and processes in a competitive setting. Figure 3 summarizes the steps involved in the development of the compensation program:

This program was developed and implemented over a two-year period with major costs including the salary of the senior-level human resources staff director and the cost of the awards for technical performance. As the program has been inaugurated only recently, data regarding total implementation costs are not available. This management approach has been used to overcome barriers in the innovation process, which inhibit access to, or retention of, technical manpower resources.

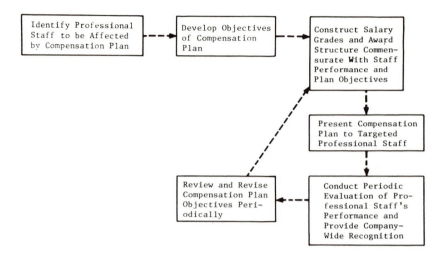

Figure 3.   Steps in developing a compensation program.

Organization

The International Harvester Company was founded more than 75 years ago with the introduction of the McCormick reaper. Since the 1930s, when the tractor was developed, major R&D efforts have been directed toward development of new technology to be used with the axial flow combine. Products manufactured by International Harvester include trucks, agricultural and industrial equipment, construction equipment, recreational vehicles, lawn and garden equipment, engines and steel. Revenues in 1980 totaled $8.39 billion, with 90,936 employees in the United States and 28 foreign countries.

The managerial structure is that of a traditional hierarchy within five broad product groupings:

- Agricultural equipment group
- Truck group
- Construction equipment group
- Components group
- Diversified group

The R&D staffs within these groups recently have made major moves to increase the level of innovation. The truck group has entered into joint venture agreements with several foreign truck-makers. It introduced six new truck models last year and became the first manufacturer of hydraulic disc brakes for medium-duty trucks and buses. The construction equipment group has introduced a number of new products, including six-wheel loaders and three-crawler tractors. During the past year, the components group has continued working toward its goal of becoming the least-cost producer of engines, castings, transmissions and hydraulics. This goal is being accomplished through investment in state-of-the-art manufacturing technology. The diversified group has applied space-age technologies (solar turbines) together with innovative product financing methods to respond to the economic pressures that accompany increased demand for products.

These five product groups require a high degree of technical perform-ance form International Harvester's R&D staff. There are approximately 800–900 professionals within this professional career category who are the target of the company's compensation program. As with the manufactur-ing side of the firm, R&D staff members are organized under a hier-archical management scheme within each product group.

Management Technique

Overview

The International Harvester Professional Compensation Program was designed to provide challenges and career opportunities for the corpora-tion's professional community, coupled with significant recognition and reward. The program has the following features:

- an eleven-level professional compensation structure, including salary, bonus, stock options and other financial rewards;
- a reward system for participation in professional societies, including cash awards for officerships, publications and advanced degrees; and
- special cash awards for outstanding technical contributions, including extra-ordinary patents.

The eleven levels or grades designated within the compensation structure are distinguished by individual performance, ability, demonstrated ac-complishments and level of responsibility. Movement among the levels depends on the individual's ability to develop these characteristics and meet the company's need.

Normally, the first three levels (P1–P3) are entry-level grades. Selection of entry level is associated with an individual's experience, education and background. The next four (P4–P7) can be compared to middle manage-ment levels and are reserved for individuals who have attained advanced professional knowledge and have made noteworthy contributions within their field. The remaining four (P8–P11) are held by professionals who have excelled in their given field and are acknowledged authorities. Usually, the individuals at these higher grade levels have accomplished work that has given them national or international prominence. These higher-level professionals have salaries and status comparable to those of upper-level managers.

To promote strong links with professional peers, International Harvester has established a Professional Awards Program designed, through the use of bonuses, to encourage individuals to participate in professional activities outside the firm. An individual receives a $500 cash award for earning an advanced degree and $300 for publishing an article in an appropriate journal. An award of $100 is given to a professional who joins an approved professional society and $200 is awarded to one assuming a position as officer within an approved professional society.

A technical contribution award is made to individuals who have made major individually identifiable contributions that directly benefit the company. The award bonus could be as high as 25% of the mid-point of the salary range for the recipient's salary grade level classification. The inventor's award is given to individuals who develop patents or patentable ideas for use within the corporation. The inventor's award bonus depends on the complexity of the new product or process and the anticipated benefit to be derived by the firm. Recently, an employee received an inventor's award of $10,000.

International Harvester also has established an incentive plan for the top six professional grades. The size of these awards varies depending on company and group profits and individual performance. Eligibility to participate in the incentive plan is based on technical performance.

Selection and Implementation of
the Management Technique

International Harvester designed this program specifically to retain top-quality engineers, scientists and researchers and to encourage them to maintain a high level of creativity. Company officials did not think any other management technique could achieve these objectives.

The International Harvester Professional Compensation Program was conceived of, and implemented by, a senior vice-president hired specifically to develop it. He has a technical background and has been involved in the management of technical research and development in other industrial applications. Throughout the program's development and introductory stages, it had the strong support of top corporate managers.

Technical staffers affected by the new compensation program were introduced to it with printed material. Employees received statements outlining the incentives. A technology advisory council, which consists of representatives of the major engineering groups, meets for two days every two months to discuss technical problems and to identify specific project objectives for the technical staff.

Through the program's structure of awards, the company gives high visibility to individuals who excel within technical fields, for example, by a formal dinner held to present the inventors' awards. During the past year, however, the company decided to delay presentation of awards because the industry is experiencing financial constraints associated with a slower economy.

Evaluation of the Management Technique

A formal evaluation of the professional compensation program at International Harvester has not been completed and one is not currently planned. The program is only two years old and is now in a state of suspended animation so it is difficult to obtain even a qualitative evaluation.

The individual awards programs have met with success in terms of acceptance by the technical staff. Long-term evaluation of the compensation program will be based on the technical staff's participation in professional activities outside the corporation, the number of patented new products and processes developed by the technical staff, the duration of technical staff employment, and a correlation between length of employment and participation in the innovation incentives program.

Although the professional compensation program at International Harvester was designed primarily to meet the career and scholarly objectives of the technical staff and to emphasize that technical contributions are highly valued by the corporation, other applications within the firm are envisioned. Disciplines such as finance and marketing possibly could offer similar compensation programs to encourage and motivate individuals.

Other organizations and institutions employing the technique are as follows:

- Xerox Corporation
- Bell Laboratories
- University of Utah
- Utility Firms

INNOVATION TRAINING—
B. F. GOODRICH COMPANY

Introduction

Background

Innovation training through organizational development has been used by the B.F. Goodrich Company, a manufacturer of tires and rubber products, to encourage innovation within the firm. Some specific areas in which the technique has been applied are financial reporting, strategic planning, organizational development, information processing and other key management activities. This management technique aims to overcome barriers associated with the communication of organizational goals related to technological innovation by those employees involved in the innovation process.

Organizational development is a recognized tool to effectively modify behavior toward a given set of objectives. To some, organizational development connotes sensitivity training, group problem-solving or team building; to others, it is associated with educational workshops conducted by the company. Extensive organizational research has shown that the operating climate of an organization is determined primarily by the cumulative efforts of managers at all levels. In addition, there is a close relationship between the management system maintained by an organization, and performance or results. Innovative behavior is more likely to be supported in a participative group management model than in one that is more authoritative.

The management technique of creating an environment or climate supportive of innovation through organizational development is based on a study of the organization to understand its beginnings, technology, resources, structure and management style. This understanding of the firm's history is necessary to evaluate previous attitudes towards innovation within the firm, as well as to predict future innovation strategy.

Organization

In 1870 the B. F. Goodrich Company founded the rubber industry; the company's principal products were fire hose and tires. The company now encompasses three major groups of products: (1) chemical products, (2) tires and related products, and (3) engineered products. The chemical products group includes plastic resins and compounds, synthetic rubbers, latexes, additives, speciality chemicals and polymers. Tires and related products are manufactured for automobiles, trucks, buses, trailers, off-the-road equipment, farm machinery and industrial equipment. The engineered products are distributed to the aerospace, mining, oil, construction and transportation industries and include such products as conveying systems, materials handling equipment, industrial rubber products, transportation products, fabricated polymers and construction and roofing products.

B. F. Goodrich Company is a member of a highly competitive industry. Although gross 1980 revenues for the firm totaled $1.29 billion, indicating only a slight drop from the previous year, pressure exists within the industry to maintain a high rate and level of innovation in the face of an apparent downturn in the economy. The management technique of training for innovation is used by the B. F. Goodrich Company not only to introduce innovative management approaches to be applied throughout the firm, but also to stimulate an environment supportive of technological innovation.

Management Technique

Overview

The successful introduction of innovation within or outside a firm, has a number of key characteristics:

- The users must believe that the innovation is better than the production process it replaces.
- It must be compatible with the potential user's values, past experiences and needs.
- It must prove useful.

An important means of fulfilling these criteria is to incorporate the need for organizational development into formal operating principles. Through the use of organizational development, it is possible for individuals to link their personal objectives with those of the firm. Individuals who have had experience in implementing organizational development plans involving the introduction of innovation may serve as instructors for related educational programs.

*Selection and Implementation of
the Management Technique*

The curriculum for the organizational development educational program used by B. F. Goodrich is accredited by a major university and alternates five one-week workshop modules with normal working assignments. These modules are as follows:

- Overview of Organization Development
- Personal Development and Innovation
- Diagnosing for Innovation
- Theory and Methods of Introducting Innovation
- Project Review and Skills Clinic

The coursework is led by university professors and leading practitioners in the behavioral sciences. Between workshop modules, participants are given assignments to enable them to experiment with innovation. When the participants finish the five-week series of courses, they begin serving as project coordinators for innovative efforts within their respective divisions. After successfully applying the techniques learned, the persons who have participated in the program are asked to play one of the following roles when innovations are introduced:

- To assist in the diagnosis of the persons or organization affected
- To help create an intent to change (if change is needed)
- To help translate this intent into action
- To stabilize the innovation and prevent discontinuance

The individuals who assume these roles must have credibility within the organization and must be recommended by their respective line or staff divisions. Transferring the skills and technology from professional consultants to line managers helps make innovation reasonably easy to understand.

Evaluation of the Management Technique

Through organizational development and personnel training, an acceptance of innovations may be achieved. Organizational development also can help management introduce innovations successfully within the organization. Organizational development efforts within a firm can result in a network of similarly trained personnel who increase the organization's capacity for forward, downward, horizontal and diagonal communications. This network provides an effective and efficient information retrieval and dissemination capability to relay information in response to the requirements of innovation. The network serves as an effective link between corporate divisions and site locations for goal setting and other work flow requirements. This network can be established to support management in the introduction of new programs. It provides a response to management at a broader base rather than within a small corporate staff. Organizational development is a method of approaching the problem of introducing innovation. An organization's ability to respond to changing needs of the future can be aided if organizational development tools are transferred from the staff functions to the line management.

Other organizations and institutions employing the technique include the following:

- Xerox Corporation
- Bell Laboratories
- University of Wisconsin
- University of Utah
- Utility Firms

INTERNAL VENTURE MANAGEMENT— MINNESOTA MINING AND MANUFACTURING COMPANY

Introduction

Background

New venture departments or groups are being established by large corporations to create centers of responsibility for the development of new business opportunities. These centers usually have an organizational structure that is isolated from other functions of the firm and they support all aspects of new product or process development. This management technique has been utilized to overcome technical and organizational barriers that exist in all phases of the innovation process. Most notably, new venture groups formed within organizations serve to overcome conservative goals and philosophies associated with the normal business operations and to investigate new, high-risk markets and products.

The concept of internal venture management, or new venture group organization, has been used increasingly by large, technology-related corporations over the past 15 years. During this time, management researchers have observed the structures of these internal organizations, their purposes and reasons for their success or failure. A firm that has successfully employed new venture group organization to identify and develop new technologies is the Minnesota Mining and Manufacturing Company—the 3M Company.

Organization

The 3M Company was established more than 50 years ago, primarily to produce products with abrasive and adhesive qualities. The company's product line has expanded from coating and bonding of abrasive and adhesive materials to include video discs, suntan lotions, a soybean-based herbicide, digital facsimile transmission, a water-immersible orthopedic casting material and many other products. There are ten product groups that are part of four major business sectors. These sectors are Industrial and Consumer, Life Sciences, Electronic and Information Technologies, and Graphic Technologies. More than 83,000 individuals are employed by 3M at more than 60 facilities in the United States and 19 sites around the world. Revenues for 1980 exceeded $5 billion. The company's current commitment to research and development represents approximately 4.5% of sales, which has resulted in more than $1 billion being spent on R&D in the past five years.

The 3M Company is unique in that managers are selected and trained almost exclusively through in-house promotions. The firm rarely recruits middle or upper management from other firms. It also has a strong incentives program designed to give a high level of recognition to individuals responsible for developing new products and processes. In a related area, more than half the manufacturing decisions of the firm have been based on the quality circle concept to improve productivity and employee involvement. These organizational characteristics provide a stable, employee-oriented environment which, coupled with an aggressive program of innovation, has made the firm a leader in the industry.

The 3M Company has utilized internal new venture management as a tool to develop new products and businesses. Typically, individuals within the firm's research fields are encouraged to introduce a new product concept. If if is supported by upper management, they can follow through with the management of the new product research and development within an organizational framework similar to that of a small profit-making firm located within 3M. Recently, 3M replaced its New Business Ventures Division with a new division called Technology Enterprises to focus on the development of families of technologies, rather than individual products.

Management Technique

Overview

Research has indicated that there are certain conditions which, if present, tend to lead to the successful development of new products and processes. These conditions center around entrepreneurship. For example, it is vital that the entrepreneur be identified and supported by the organization at an early stage in the development of the product or process. The corporation must formally recognize the influence of the innovator and give the individual license to perform the development project. The entrepreneur also should be afforded direct sponsorship by a representative of upper management to ensure that the project is located at a strategically viable level within the organization and that the entrepreneur receives discretionary powers that will permit unhampered development of the project.

The formation of new venture groups within a large corporation provides a successful entrepreneurial environment. A new venture group offers an independent setting for the entrepreneur to deal with the instability and uncertainty inherent in new business development. By establishing separate organizational entities within the firm to deal with the vagaries of new product development, the parent organization can maintain the regular and predictable climate necessary for the ongoing business while expanding new product avenues.

There are no set structures for new venture groups. They may consist of a small number of individuals or be a large group, or they may deal with individual products or broad business concepts. There are, however, some general tendencies or characteristics of internal new venture groups dealing with the development of new technologies.

The groups are established as profit centers separate from the traditional financial structure of the organization. Decisions to establish and support new venture groups at given budget levels are made by upper management. Each new venture group is led by an entrepreneur, called a "product champion," who usually is supported by a varied staff of professionals drawn from all segments of the firm. In addition to project engineers and scientists, the new venture group will include marketing representatives and strategic planners to assist in the mapping of the product life cycle.

New product development projects managed by new venture groups commonly pass through four distinct phases from generation to commercialization. During the first phase, technical and commercial evaluations of the new product or process concept are conducted. Often this first phase of new product development is performed by the innovator and management within the traditional organizational framework. Once the idea has proven to have economic merit, the second phase of new product development is initiated. During this phase, the new venture team is formalized and a detailed plan for development is constructed. While the business plan is being developed, product profitability is analyzed. (Firms that support multiple new venture groups/projects review new product ideas and their respective anticipated return on investment in relation to the portfolio of projects offering different risks and rewards, as well as differing maturity dates.) The third phase of the process of characterized by major product and market development activities. During phase four, the new product is marketed through a new or existing corporate division, or is licensed or spun off into a free-standing enterprise.

Selection and Implementation of the Management Technique

The Technology Enterprises Division at 3M is a permanent entity responsible for guiding new product ideas through the four phases of development outlined above. This division draws ideas and personnel from the Central Research Laboratories at the company headquarters. As an impetus to encourage innovation, top management has maintained a goal of deriving 25% of sales from products that did not exist five years ago.

The 3M Company, and other large corporations that have adopted the internal venture management approach, have assumed general guidelines for forming and operating the groups:

- The new venture group should be formalized through printed documents defining its scope, purpose, constraints and activities.
- All levels of management of the firm should be educated as to the purpose and goals of the new venture group.
- An evaluation method and timetable must be established for the consideration and review of new product ideas.
- It is advisable to include a balanced mix of new ventures in the corporate investment portfolio.
- Although the initial stages of new venture development (idea generation and project evaluation) may be covered by overhead funds distinguished from new venture project budgets, it is important that an individual budget be assigned to each new venture.
- An open communications link must be established between the new venture and top management to facilitate the review of the project's progress and provision of technical advice regarding the design and implementation of business plans and projections.
- To ensure successful diffusion of the newly developed technology, the new ventures should not be moved to another part of the organization until the project revenues are well above the break-even point.

As implied by these organizational and operational guidelines, the successful conduct of a new venture group's activities within a large corporation is highly dependent on the interaction of the group with other elements within the corporation and the operation of the group as a small business.

Evaluation of the Management Technique

Necessarily, the success of new ventures initiated by a firm, whether developed internally by a new venture group or by the traditional organizational group, or externally by another firm, is measured by return on investment. As noted above, the business plans prepared for the new ventures include projections of anticipated returns or profits. These projections are constantly modified as the project proceeds to reflect a predicted strength or weakness of the product in the market.

A major problem associated with internal venture management is the difficulty in setting realistic budgets and benchmarks against which performance can be measured. This problem is caused by the corporation's reliance on traditional budgeting and planning techniques, which do not easily apply to innovation activities. More adaptive, responsive control measures must be applied to effect more accurate predictions and measures of new venture project success.

The 3M Company, Ralston Purina and Exxon Corporation have used internal venture management techniques successfully primariy as a result of the long-term corporate commitment to the concept. Corporations involved in venture management have been the innovation leaders in their respective industries. The 3M Company's success in using internal venture management may be attributable to the relative stability of the firm and its apparent commitment to maintaining its innovative spirit.

Other organizations and institutions employing the technique are as follows:

- International Harvester Company
- B.F. Goodrich Company
- International Telephone and Telegraph Corporation
- McDonald's Corporation
- Bell Laboratories
- University of Utah

PRODUCT CHAMPION—DONALDSON, LUFKIN & JENRETTE, INC.

Introduction

Background

A "product champion" is an individual who plays a key role in selling an idea to management, maintaining management's interest in the project and bearing the risks associated with the development of a new venture. This management technique has been used by a wide range of firms as a formal means of managing internal new ventures. The product champion also is used in the management of new ventures outside the firm. A particular example of the product champion role can be observed in the description of two other management techniques included in this study: (1) external venture management, as seen in the Sprout Group of Donaldson, Lufkin & Jenrette, Inc., and (2) internal venture management, as used in the Minnesota Mining and Manufacturing Co.

Organization

The product champion often performs his or her role in the organizational setting of a special group within a firm that has been established to develop a new product or process. This group is often formed as a result of the efforts of the product champion and, therefore, individuals who participate in a new venture development project with a product champion will necessarily be supportive of the champion's ideas and plan. Characteristically, the product champion is somewhat isolated from other functions within the organization by virtue of the assumption of risks associated with the new product or process.

Management Technique

Overview

The new venture manager or product champion faces the risk of failing to produce results, the effects of unknown and unforeseen variables, the possibility of being isolated from the corporate mainstream and the lack of support (financial, political or technical) from within the organization and top management.

The literature describing innovation in large U.S. corporations generally describes the role of a product champion as overseer and director of new product groups, whose purpose is to overcome the status quo attitude so often prevalent in large firms. Systems and procedures have been established in large corporations that quite directly inhibit the success of new product innovation. Studies conducted to characterize the product champion have revealed several key qualities that can be attributed to the individual in that role.

The product champion is usually someone who has a sound technical understanding of the product and can assess realistically the new product's limitations and advantags. This ability to evaluate the future of the new product often is based heavily on technical intuitiveness with very few supportive data. The product champion also must be able to give a realistic appraisal of the new product's market potential to facilitate the development of a suitable marketing strategy. Most significantly, the product champion must possess the aggressiveness to get the work done and decisions made, as well as the political astuteness to identify power centers and communicate successfully with individuals at different levels within the organization.

Selection and Implementation of the Management Technique

As a key manager throughout the process of innovation, the product champion must be able to assume multiple roles. To accomplish this, he must have a clear understanding of the organization's needs and goals and be able to identify organizational resources and constraints that will affect the new product's success. During the management of a new product, a product champion must serve as stimulator, initiator and legitimizer. As stimulator, the product champion must start the process of idea generation to encourage the development, identification and selection of new product projects. As initiator, the product champion translates the idea into a plan of action and presents the plan to key individuals within the organization. As legitimizer, he has the social power to sanction the idea. The extent to which the product champion may perform as a legitimizer depends on the breadth of his experience, the size and structure of the organization, and the type of new products or processes being developed.

Evaluation of the Management Technique

As mentioned above, the role played by the product champion in the successful development of new products and processes is critical in large organizations where conformation to the norm is expected and innovation is almost a forced process. The success of the product champion's performance is measured, quite obviously, by the relative success of the new venture. More indirectly, however, the success of the product champion may be measured in terms of his ability to mitigate the risks associated with developing new products and processes, as well as his talent for performing specific functions related to his role.

In new venture operations that have utilized product champions, successful individuals have emerged naturally to assume the new product management responsibilities, rather than being selected through traditional methods of promotions and job reclassification. Product champions may be from the ranks of top- or mid-level management, or from the technical project level. Although project-related problems addressed by product champions may vary in size or scope, depending on the industry, similar personality and professional characteristics are common to individuals willing and able to assume the role.

Other organizations and institutions employing the technique include the following:

- International Harvester Company
- 3M Company
- Xerox Corporation
- International Telephone and Telegraph Corporation
- Bell Laboratories
- The MITRE Corporation
- University of Wisconsin
- National Aeronautics and Space Administration
- University of Utah

PROJECT PERFORMANCE MEASUREMENT— A TECHNICAL MANUFACTURING FIRM

Introduction

Background

As part of a major effort to increase the effectiveness of project groups in a high-technology matrix organization, a major U.S. company* engaged in the manufacture and marketing of technologically sophisticated products for industrial and consumer use has adopted a procedure for measuring project performance. This procedure was selected to increase the information feedback between managers and technical staff

---

*Because of the proprietary nature of the circumstances surrounding the application of this procedure, the firm cannot be identified.

to enable managers to evaluate decisions regarding project selection and project monitoring in relation to project outcome. The major barriers affected by the project performance measurement approach are those organizational and structural barriers that inhibit the flow of informtion regarding project evaluation.

The project performance measurement procedure was developed over a one-month period by an independent management consultant.* This procedure consists of several distinct stags in its development and implementation. As shown in Figure 4, the initial five stages of the procedure require the preparation of a questionnaire to assess project performance. This questionnaire is then used to survey participants in technical projects, and an analysis of the data collected permits a standard profile of the project to be drawn and compared with that of other technical projects within the firm. The comparative data provided by the analysis of projects using this procedure enables managers to identify more readily weak areas within each project's organizational framework.

Organization

The project performance measurement procedure was developed at a firm that employs approximately 30,000 employees; total revenues for 1980 were $1.4 billion. The company's major products include glass and

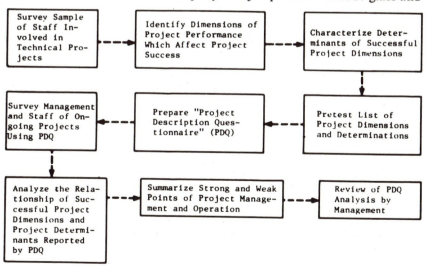

Figure 4.   Project performance measurement procedure.

---

*Thomas DeCotiis, University of South Carolina, Columbia.

ceramic materials, components and systems for industry and scientific research.

The R&D staff of the firm is approximately 800, 400 of whom are scientists and engineers. Following matrix organizational lines, the technical staff is assigned to discipline-oriented groups such as chemical and physical research, as well as to project-oriented teams that cut across several disciplines. The project-oriented teams are often responsible for following a new product or process from the initial stages of proposal selection through basic and applied research, development and manufacture.

Management Technique

Overview

The criteria on which projects are evaluated using the project performance measurement approach were derived from a survey of 20 staff members who represented a cross section of the company's functional and technical skills and who had been involved in one or more new product development projects. The interviews were structured to focus on the identification of dimensions of project performance that could be evaluated and the characterization of the determinants of the project performance. It was intended that the information obtained through the project evaluation permit project managers to plan changes in unsatisfactory project operations and to facilitate better project management in the future.

From the survey of project participants in the case study firm, five dimensions of project performance were identified. These dimensions are summarized below:

- **Manufacturability and business performance**—measurement of the ability to manufacture the product and obtain a favorable financial return.
- **Technical performance**—measurement of the ability of the project team to generate the required technical data and meet technical specifications.
- **Efficiency**—measurement of project operation within cost, time and productivity goals.
- **Personal growth experience**—measurement of the effect of the project on the participants' professional and personal development.
- **Technological innovativeness**—measurement of project results in terms of technological advances.

An important aspect revealed in the identification of these dimensions is that a project can be evaluated as a success or failure depending on the orientation of the evaluator—management or scientific. The survey data collected by the consultant indicates that project participants viewed high performance in the manufacturing and business performance dimension as the best indicator of project success, whereas high efficiency and technical performance were viewed as the major contributors to success. Although technical innovativeness and personal growth experience are important, they are not perceived to be strongly related to project success.

The survey also identified specific determinants of project performance, which can be divided into three groups concerning (1) external environment, (2) relationship between the project and the functional organization, and (3) internal project operations. The specific determinants included within each of these categories are presented as follows [5]:

External Environment

- **Management support**—involves the attitude of the management of the project, as shown by project priority, resource availability, involvement of top management and project personnel in decision-making, and reward policies.
- **Interorganizational relations**—describes the nature of the relationships among the sponsoring divisions concerning project goals and specifications, authority relationships, problem solving and project financing.
- **Sponsor relations**—involves the degree of support and assistance provided to the project by the sponsoring division(s) and the nature of the interaction between the sponsoring division and the technical and manufacturing staffs.
- **Transfer management**—describes the transfer process, including such aspects as the locus of responsibility for timeliness, planning, priority assignment and the nature of the interaction among the involved divisions.
- **Planning for and stability of specifications and designs**—describes the extent to which technical specifications and process designs are planned and clearly stated in advance of various project phases and the extent to which they are modified as the project unfolds.

Relationship Between the Project and Functional Organization

- **Project leader, functional manager relations**—describes the nature of the interaction between managers of the discipline-oriented resource groups and the project leader, with particular emphasis on the coordination of decisions, the delegation of authority and the degree of latitude allowed the project manager to carry out project tasks without interference.
- **Clarity of the Project Leader Role**—involves the extent to which the project leader and his authority are defined and clarified and the amount of control the project manager exercises over project personnel.

Internal Project Operations

- **Project members' skills and cooperations**—involves the extent to which the skills and experience needed by project personnel are planned for in advance, successfully obtained and subsequently integrated into a coordinated and cooperative effort toward project goals.

- **Communication, decision-making and personnel utilization**—pertains to internal operations of the project, with particular emphasis on the flow of communications, methods used to make critical decisions and the ways in which project personnel are utilized.
- **Planning and scheduling**—describes the extent to which project goals are broken down into specific tasks. Task responsibilities are clearly assigned and procedures for accomplishing the tasks are worked out carefully.
- **Control procedures**—describes the extent to which formal control techniques are used and the importance assigned to meeting budgets and deadlines.
- **Leadership**—includes the formally designated leader's knowledge and competence and the various activities engaged in by the leader to get decisions made, problems and conflicts resolved and needed information communicated.

To ascertain the relationships between the project performance determinants and the performance dimensions, the dimensions were stepwise regressed on the determinants. Highlights of this analysis indicate that the manufacturability and business performance dimension is most affected by the transfer management determinant (a part of the external environment). Other important determinants of manufacturability and business performance are the planning and scheduling activities associated with the project, interorganizational relations and clarity of the project leader's role. The technical performance dimension is most highly dependent on the external environment of the project. Project efficiency involves determinants including the quality of transfer management, the planning and stability of specifications and designs, and project members' skills and cooperation. The personal growth experience dimension depends on communications, decision-making skills and personnel utilization, as well as good interorganizational skills and cooperation. Technological innovativeness, although not strongly affected by any one determinant, correlates with the level of communications, decision-making and personnel utilization within the project.

Selection and Implementation of the Management Technique

The company's decision to adopt this method of project performance measurement was the result of several factors. Managers of the technical R&D projects within the firm were unable to predict project success accurately. They could not identify specific factors that generally would be indicative of successful performance. Managers also were unable to select personnel with the appropriate qualifications to participate in the technical projects and programs. At the time, this approach to project evaluation was considered and ultimately implemented a consultant who was working with the firm to address the related problem of increasing the effectiveness of project groups in a high-technology matrix organization.

Results of the survey conducted by the consultant were pretested with a sample of experienced project participants and used to prepare a "Project Description Questionnaire" (PDQ). The questionnaire was used to evaluate the performance of specific projects within the technical staff division. The data derived from the survey were analyzed to characterize ongoing projects using a common framework.

Evaluation of the Management Technique

According to feedback received by the management consultant from the project managers in the firm, the PDQ is useful for evaluating a project's performance after its completion when considering manufacturability and business performance. Concentrating on the remaining four dimensions included in the project performance measurement, ongoing projects can be evaluated. The modeling of a project's organization and behavior using the PDQ provide useful insights in providing the organizational climate and structure supportive of successful projects.

Future use of this project performance evaluation approach probably will incorporate a modified Delphi method for obtaining revised project dimensions and determinants. It is anticipated that through the use of the Delphi method and inclusion of corporate strategies, project goals may be linked more closely with corporate goals and objectives.

This technique also is used by the University of Utah.

QUALITY CIRCLES—FORD MOTOR COMPANY

Introduction

Background

The Ford Motor Company, one of the leaders in the American auto industry, is concerned with innovation because the automotive market is highly competitive and calls for the creation of more efficient, attractive, convenient or otherwise "improved" models every year. This high degree of product change has demanded the development of a process that yields a constant stream of new project ideas. In recent years, however, traditional patterns of innovation in the industry have been altered by increased government regulation, rapidly rising fuel prices and aggressive foreign competition. As a result, managers at Ford, as well as other American automakers, have had to evaluate and develop new approaches to the innovation process.

A look at managerial principles used by the major competitors of American auto manufacturers, the Japanese, indicates some differences in approach to technological and process innovation. One area in which the Japanese have excelled is that of quality control through reactive innovation, the development of new products in response to identified needs—market pull. In an effort to improve its quality control, Ford, as well as many other auto manufacturers and industries in the U.S., decided to adopt an approach used by the Japanese, which draws workers into the decision-making process. This approach is the quality circle concept, which was developed primarily to involve employees in periodic problem-solving and to make them aware and supportive of the goal of higher productivity. Figure 5 shows a simplified sequence of events that may be followed in the establishment of quality circles. As these circles have been established in many firms, experience has shown that they offer a good environment for the development of new products and processes in manufacturing and are supportive of higher productivity levels.

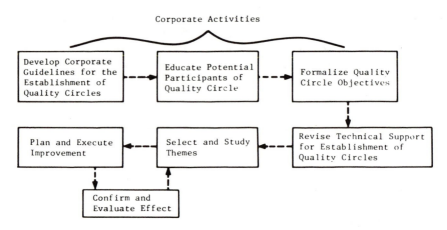

Figure 5.   Steps in establishing a quality circle.

The intent of this management approach is to remove communications barriers that inhibit the transfer of corporate goals and ideology to the workers and to facilitate communication of innovative ideas regarding improvements in products and processes from the workers to decision-makers. Quality circles provide a formal setting in which communication channels can be strengthened and participation in the innovation process enhanced.

Organization

The Ford Motor Company has approximately 500,000 employees located at more than 90 sites around the world. In 1980 Ford's automotive sales totaled $33.30 billion; sales for other products and services generated $3.78 billion. Ford has more than 13,500 dealers who sell Ford-built cars, trucks and tractors in nearly 200 countries and territories throughout the world. Other business interests held by Ford include a conglomerate of six manufacturing divisions, a worldwide tractor organization and Ford Aerospace and Communications Corporation, the subsidiary responsible for INTELSAT V, a series of communications satellites. Ford also owns finance and insurance subsidiaries, which provide market support for Ford's product dealers. The Ford Motor Land Development Corporation, another subsidiary, is involved in the development of a commercial and residential community complex.

The organization of Ford's manufacturing operations is hierarchical, with approximately seven tiers of management between the production line and corporate staff levels. Most manufacturers within the automotive industry are structured similarly.

Management Technique

Overview

Quality circles, a term used to describe an organizational structure designed to influence factors of product quality, were developed by the Japanese. In the 1950s, under the guidance of W. Edwards Deming, an American statistical consultant, Japan's industrialists became interested in adopting a new approach to quality measures. Known as the founder of the third wave of the Industrial Revolution, Deming's statistical analysis methods enabled management to evaluate problems and interpret signals of needed changes. An apparent bonus of the quality circle system has been that it elicits valuable ideas from workers—encouraging innovation and inventiveness.

General characteristics of the quality circle concept are summarized below:

1. Participation in the quality control circle must be voluntary.
2. Participants must be trained in statistical analysis, group dynamics and problem-solving techniques.
3. Members of the quality control circle must be able to select the problems to be addressed and, where possible, implement the solutions and monitor the results.

4. The quality control circle must be permitted and encouraged to meet during company-paid time.

Some of the specific problem analysis tools used by quality control circles are brainstorming, logging of problems, graphic illustration of problem log data, cause and effect diagrams and other statistical illustrations such as histograms, scatter diagrams, graph and control charts, and stratification sampling.

Selection and Implementation of the Management Technique

Ford's decision to implement the quality circle approach to improve productivity and encourage innovation was based primarily on three factors. First, other members of the industry had successfully implemented quality control programs similar to those employed by the Japanese. Quality circles were viewed by employers as an environmental attribute of the job; therefore, it was necessary for Ford to provide equal opportunities for employees to participate in problem-solving sessions related to work quality. A second reason for selecting this management approach arose from employee opinion surveys conducted at Ford. They indicated that employees were more productive and tended to be more innovative when they sensed greater involvement in the firm's operations and greater control over their work enivronment. Finally, Ford chose to stimulate innovation through organizational behavior modification. In the automotive industry, production technology is somewhat fixed, but there are numerous opportunities to affect the organization and behavior of human resources and thus improve product quality.

Employee involvement through quality circles has been implemented at 50 sites during the past two years, including 250–300 groups and approximately 4500 employees. The ultimate goal of the program is to develop 3000 groups at 90 locations, each involving 30–40 employees. Groups consist primarily of hourly wage employees and their supervisors. Most problems addressed by these groups involve the assessment of the group activity with regard to productivity. When change is indicated, the group attempts to effect the required changes through behavior modification.

The employee involvement program at Ford was established in concert with the United Auto Workers (UAW) union to indicate joint corporate/union involvement and support. Three different approaches have been used:

1. **Employee Involvement Problem-Solving Groups**. Small groups of employees meet with their supervisor to identify and solve quality and other work-related problems. These groups are provided training that includes problem identification and causal analysis.
2. **Employee Involvement Quality Circles**. These groups are provided training that includes cause and effect diagrams, graphs and Pareto diagrams, as well as basic statistics.
3. **Employee Involvement Team Building**. This brings together employees for the purpose of accomplishing a common objective focusing on the interpersonal aspects of problem-solving and the accomplishment of group objectives.

Through the use of guidelines and operating committees, the basic awareness of the concept of employee involvement was presented to the employees. At each site, the following common schedule of events generally marked the development of an employee involvement program:

- Formation of a union/management steering committee
- Diagnosis of organizational problems
- Selection of plant site
- Training of participants
- Conduct of the program
- Refinement of the program
- Evaluation of the program

Evaluation of the Management Technique

Although a formal evaluation of the Employee Involvement Program has not been conducted by Ford's management, some obvious positive changes have been recognized. For example, there appears to be improved cooperation and more open communication between employees and supervisors. There also appears to be stronger mutual understanding and respect among employees and supervisors. It is expected that the individual groups eventually will be evaluated in terms of number of grievances exposed, absenteeism, level of supervisor/employee and company/union communications and amount of scrap produced during production.

Ford's corporate managers who have been involved in the implementation of the program have noted that the success experienced with the program may be attributed to two actions. First, prior to initiation of training programs the president of Ford issued a joint letter of understanding with the union management outlining common goals: to "make work a more satisfying experience, improve the overall work environment, and enhance creativity, contribute to improvement in the workplace, and help achieve quality, efficiency, and reduced absenteeism."

The other action that has influenced the successful introduction of the employee involvement program has been the use of outside consultants to facilitate the development of groups. Consultants act as part of the Ford Employee Involvement Program to assist with diagnosis as well as development and implementation of pilot projects, to serve as a go-between for different groups, or to develop training program materials.

Corporate management has predicted that the target level of program participation will be reached over a 10-year period. Then the program will be evaluated to determine the degree to which it has been modified from its original format and internalized by the organizational structure. Results of the program, in terms of innovativeness, will depend on the demands for innovation occurring naturally in the competitive market, as well as by the environment within the firm and the industry.

Approximately 750 U.S. companies and government bodies have employed quality circles. The U.S. approach to the quality circle concept differs from the Japanese model in that production-related decisions are generally made at the operations level in the United States and at higher levels of management in Japan. The American approach also differs from the comparable European system of codetermination, in which worker representatives sit on corporate advisory boards but do not assume direct responsibility for adopting innovation within the production process.

Some of the other American firms that have implemented the quality control circle approach have been able to realize rates of return on invested operating funds of up to 17 to 1. At Lockheed's missiles system division, 33 circles were established over a five-year period at a cost of approximately $700,000. During that same period, cost savings of more than $5 million were realized. Westinghouse, another major U.S. manufacturer, operates more than 25 circles at more than 50 locations in the U.S. Just one of these circles (in the purchasing department at the headquarters of the defense division) saved more than $600,000 for the company.

A review of quality circle systems in the U.S. reveals that the first circle can be established and operating within weeks. It generally costs between $8000 and $15,000 to launch a quality circle program using an outside consultant. According to general estimates made by industry analysts, approximately 80% of proposed solutions to problems developed by the circles in the U.S. have been implemented.

Other organizations that employ this technique include International Harvester Company and a technical manufacturing firm.

RESEARCH PLANNING FRAME—
XEROX CORPORATION

Introduction

Background

The Xerox Corporation employs a management technique called Research Planning Frame, which sets forth goals, activities and schedules for the period during which a new technology is identified, researched and developed. This management approach was designed to ensure accurate long-term planning of new products and corporate goals related to innovation. Specifically, the research planning frame strives to involve persons who know of the corporate philosophy and industry trends in the project selection—the first phase of the innovation process.

Figure 6 outlines the steps taken during application of the research planning frame.

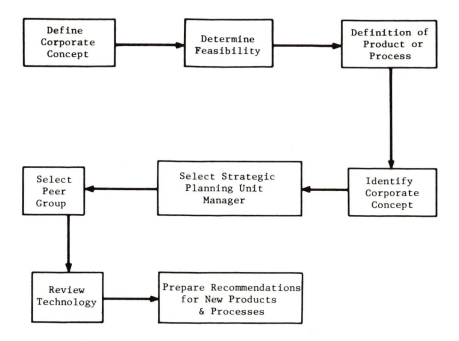

Figure 6.   Steps in applying the research planning frame.

Organization

The Xerox Corporation is a multinational firm involved in information processing, reprographics and office systems. Revenues for 1980 totaled $8.2 billion, representing a 17% increase over 1979. The company had 120,480 employees at the end of 1980, representing a growth of 4% since the end of 1979.

The company supports a research organization that provides the environment for new product definition and development. Corporate research is conducted at three sites and four research center organizations located at six separate geographic locations. The primary function of the centers is to generate a continuous flow of ideas, inventions and prototypes to provide the functions and services supportive of higher productivity in the business office. The research programs of these centers therefore must encompass basic research in services relevant to technologies used by the corporation; applied research directed toward new or improved technological capabilities; and selected exploratory development of research prototypes as forerunners to new products and services.

Individually, the research laboratories focus on the following research areas:

- Webster Research Center—pictorial and electronic informational representation
- Palo Alto Research Center—imaging and printing technologies
- Xerox Research Center of Canada—imaging science and information transfer

Xerox utilizes the standard technology transfer patterns between the research and engineering organizations within the company to encourage the development of new products and processes, as well as to create an environment in which individuals can launch themselves into careers within the corporation.

Management Technique

Overview

The Xerox Corporation has developed a long-range planning system in an effort to anticipate new products and processes that need to be developed 10 years from now. The main reason for formalizing the long-range planning and investment process is the nature of the innovation process within the field of xerography. It generally takes 5–20 years to complete the process of idea generation, feasibility testing, development, design, engineering and production. Each stage in the process may last as long as five years. Xerox's is defined as that management planning time frame during which a new product or process can be researched and developed to the point at which a business impact can be felt.

An important part of the planning process is to develop a set of company goals, specific down to the level of kinds of products and services to be provided. This process involves senior management and the establishment of special staff groups and program teams that meet periodically to create and refine the corporate goals and new product and service concepts.

The basis of the research planning frame is to define a market opportunity. During the conceptualization phase, new products or processes are aligned with marketing goals. Corporate participants representing all operating units within Xerox review the concepts to assess their market potential. During the feasibility stage, refinements are made on production costs and ultimate service costs. Based on pricing levels, market opportunity projections are made and market penetration is estimated. The final definition phase is that period when the product or process is finalized.

Selection and Implementation of the Management Technique

Xerox decided to implement the Research Planning Frame because of the innovation climate within the industry and the speed with which new products and processes were being developed. Implementation of the long-range corporate planning process required corporate leaders to answer the following questions:

1. In what business do we want to be involved?
2. What technologies are important?
3. What are the technology trends?
4. What is the company doing about them?

The answers to these questions formed the corporate game plan and required approximately two years of planning. The business concept that evolved was the "office of the future." Strategic planning units consisting primarily of research and engineering staffs were established to define new products and processes in support of this concept. Each unit is headed by a technology portfolio manager who is a technical specialist within a major scientific area. Eight technical managers are named to serve one year each in this planning capacity and usually are drawn from the research centers. Examples of the technical areas of expertise are distributed computing, programming, languages, optics, image processing and integrated systems design.

The managers selected from the research centers establish a working group of peers to develop new product and process scenarios for the next 10–20 and 20–30 years. The groups must prepare working reports outlining the idea structures and plans, as well as providing input to the long-range budgeting and marketing plans.

Evaluation of the Management Technique

The Research Planning Frame has been in operation for approximately five years. In general, the process has proven to be succesful as the budget decisions for corporate investment planning rely heavily on the recommendations of the strategic planning units. These units are said to be successful because:

1. the managerial role rotates, causing an infusion of new ideas every year; and
2. the technical professionals who participate in the planning groups are satisfied to assume this temporary post as it is prestigious and yet permits the professional to continue a regular work schedule.

No other organizations employ this technique.

TEMPORARY GROUPS—
INTERNATIONAL TELEPHONE
AND TELEGRAPH CORPORATION

Introduction

Background

The International Telephone and Telegraph (ITT) Corporation has employed a management technique to pierce barriers that inhibit the flow of communications among groups of employees and particular managers who are dispersed geographically. The major objective of improving communications was to develop a new generation of products. The technique involves the establishment of temporary groups, who apply problem-solving skills to identify new products and processes. Unique characteristics of this management approach for defining innovative ideas are that several levels of management are represented in the group, and

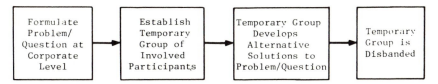

Figure 7.   Steps taken by ITT participants in the temporary group process.

the group's existence is limited in time. Figure 7 summarizes the steps taken by ITT participants in the temporary group process:

Organization

The ITT Corporation is a multinational, multiproduct firm with offices in several locations. The company is the world's second largest telecommunications equipment producer, behind American Telephone and Telegraph Company's Western Electric subsidiary. In addition, ITT oversees or invests in other businesses such as automotive and industrial products, food products and consumer appliances, processing timber and minerals, production of coal and oil, financial services and insurance. The company has more than 368,000 employees, and revenues for 1980 totaled $12.5 billion.

As in most large corporations, the division of work and functional centers at ITT are distributed geographically. The patterned hierarchical levels established within the organization often cause inefficiencies in the free flow of information.

Management Technique

Overview

ITT has utilized temporary groups primarily to stimulate innovation. Although temporary groups may be classified as committees, working groups or task forces, the distinct characteristic is that the group is temporary. Its work is bound by given time limits. The groups at ITT consist of representatives from several different hierarchical levels representing all or several sites. At the onset of the group's formation, a very explicit goal and associated tasks are assigned to the group by the corporate staff. The group is considered a "peer group," established with diagonal and random communications channels through the organizational structure, which contribute to resolution of the temporary group's task.

*Selection and Implementation of
the Management Technique*

The intitial objective in establishing temporary groups at ITT was to bring together company representatives from six countries to familiarize them with problems in each country's market. The group also was to evaluate a technology area or specific type of product to assess how it should be developed, manufactured and marketed most efficiently and with international appeal.

Twelve product-oriented temporary groups were established. Each consisted of a corporate technical staff member acting as chairperson, a corporate marketing individual, a European marketing representative and technical staff member, development heads of specific factories who were working to develop the product, heads of factories that were going to copy the work, and other selected representatives from these groups (e.g., the marketing representative from Germany).

Another temporary group, the ITT Research Council, was established later to advise the company regarding future investments in the corporate research and advanced technology program. This council which still exists, includes five U.S. group laboratory directors, two U.S. group product planning directors, three Common Market group technical directors, two department heads in two European laboratories, and one divisional chief engineer. These individuals were separated into three classes serving three-year terms. Other appointments to the Council included two product group managers, two product line managers, three technical directors, two group general managers, and a market development director. The counsil also includes the general technical director of ITT, his two deputies and the research director.

The council has provided an opportunity for a line of communications to be established from the corporate level downward to enable innovation goals to be passed through several management levels. The council setting also stimulates technical feedback from the grass roots level to top management regarding the needs of the business. As a result of the establishment of this council and the bottom-to-top flow of information, the corporate research and advanced technology program has been modified to increase the program's acceptance and value.

Evaluation of the Management Technique

The use of temporary groups in a large corporate setting has been directed toward reducing resistance to change by individuals in authority and removing the demotivating impact of the large centralized corporate structure, as well as its effect on "innovators" who are not interested or involved in long-range planning. Generally, groups can solve problems more effectively than individuals because greater knowledge and information can be brought forth. This allows a greater number of approaches to problem-solving to be defined. Group sessions also increase participation in the decision-making process, which increases general acceptance and understanding of the outcome or decision. Some drawbacks that are noted in group problem-solving sessions are (1) the negative impact of social pressure for conformity; (2) the tendency of the group to accept the first consensus (not always the best idea); and (3) the tendency of some individuals to dominate group meetings.

The temporary groups that were formed to devise and launch new international products accomplished the objective of improving the flow of new ideas and corporate innovation goals throughout the corporate management chain. Most accomplished their respective product development goals within 3–3½ years. Since the 12 original temporary groups were established, at least two additional groups have been initiated; both failed to meet the objective of stimulating innovation. The first failed because the question serving as the focal point of discussion was not phrased in a manner that permitted a solution. The second group failed as it became a "permanent" committee, merely superimposing another level on the corporate management structure.

Based on this experience, a corporate technical advisor has said that the success of the temporary group concept depends on three factors: (1) the selection of a leader who can effectively run a meeting of 50 individuals; (2) the accurate definition of the problem to be addressed; and (3) the establishment of the group on an ad hoc basis, which, however, is not allowed to become permanent.

Other organizations employing this technique are as follows:

- B.F. Goodrich Company
- 3M Company
- Xerox Corporation
- Argonne National Laboratories

EXTERNAL VENTURE MANAGEMENT—
DONALDSON, LUFKIN & JENRETTE, INC.

Introduction

Background

External venture management is a concept used to describe what occurs when an investment group provides technical management and financial support to a new or existing firm. As implied, the major innovation process barriers affected by this technique are primarily the lack of expertise in the areas of marketing and long-range planning. Without these resources, a small firm trying to develop new products cannot participate in the competitive market.

Venture management and capital firms have increased in number during the past 10 years to accommodate the growth of new firms and, more importantly, to fill the resource cap in the field. During the past three years, there has been a substantial flow of captial in the venture market. In 1980 more than $900 million was raised by limited partnerships, almost double the amount raised for the previous two years. Capital resources available for venture investment at the end of 1980 amounted to an estimated $4 billion, substantially more than the static $2.5 billion available during the 1969–77 period. These funds are controlled by different organizations and institutions, as indicated in Figure 8. With the increase in funds available, there is heightened competition for new investments. As a result, investment costs are rising rapidly.

The Sprout Group is a venture capital partnership under the direction of Donaldson, Lufkin & Jenrette, Inc., a major full-service banking firm with headquarters in New York. The Sprout Group is one of the largest venture capital operations in the United States. Since 1970 it has put more than $75 million into more than 70 small firms. During the first few years of operation, the Sprout Group, like many other venture capital firms, invested in the development of new products—the inventor and his idea. The current trend in investment of venture capital funds, however, is through the leveraged buy-out, which is, in essence, an investment in mature businesses that become available as conglomerates divest themselves of subsidiaries. Although the Sprout Group has invested the majority of its funds in the leveraged buy-out options, the company has received its greatest return on investment from startup firms, or those developing new technology.

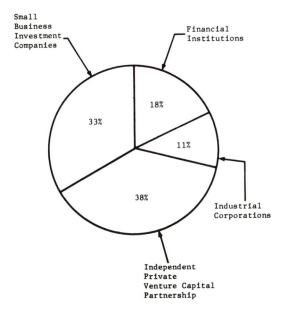

Figure 8.   Types of organizations and institutions that control venture capital.

Organization

Donaldson, Lufkin & Jenrette, Inc. was founded in 1959 to provide a full line of investment brokerage services. It is the largest nonbank investment manager in the United States, with more than $12 billion in assets. The six operating division of Donaldson, Lufkin & Jenrette, Inc. are:

1. Pershing, a network of major regional securities firms
2. Equities Division, a broker-dealer-underwriter of equity securities
3. Fixed Income Division, a securities investment arm
4. Investment Banking Division, which provides captial-raising and other financial services for corporations and public entities.
5. Investment Management Division, the largest nonbank manager of pension and retirement funds
6. Real Estate, a real estate investment subsidiary

The Sprout Group, a part of the Investment Banking Division, is organized as limited partnerships. The limited partners include banks, insurance companies, corporate pension funds, endowment funds and wealthy individuals who each have invested $500,000 to $4 million. There are approximately 12 staff members, each responsible for managing several projects or investments.

Management Technique

Overview

The Sprout staff selects firms to be included in the group's investment portfolio with hopes of purchasing a controlling interest in the firm, and aim for a pretax return on investment of 35% Most of the firms considered by the group are recommended by someone in the company who has received information from a known source. Staff members have expertise in a wide range of areas, which has been a critically important factor in the review of potential investments. During the past year, the group invested in 9 of the 75 proposals submitted from within the firm. Another 525 were received by the group from outside; only two of these were accepted.

A proposed project is reviewed by the staff and a special advisory committee, representing the limited partners. One staff member serves as project manager and follows the proposed project through review, investigating it, recommending whether and how much Sprout should invest, structuring the financing, and serving on its board. The project manager also prepares a business plan with the portfolio company's management, setting performance goals. Based on these goals, management's compensation, including equity and cash bonuses, is developed. It is then the project manager's responsibility to oversee the management of the new project once approved and realign the goals set for the project, as well as the portfolio.

Selection and Implementation of the Management Technique

One of the earlier firms developed by the Sprout Group was a new company called Envirotech. Sprout provided investment funds to a San Francisco industrialist who had the idea of creating a corporation that could produce a full range of air and water pollution control products. The company was formed by purchasing two subsidiaries of larger companies. (Within six years, Sprout sold Envirotech and realized a 400% profit.)

The management of the new venture projects by the Sprout project officers is considered the key to the successful operation of Sprout. The project officers are considered a more valuable resource than the venture capital.

A list of the firms that have been in the Sprout Group's portfolio since 1969 are listed in Tables XIX, XX and XXI. The firms are grouped into

Table XIX.   Emerging Growth Companies

Name of Company	Industry
Advanced Medical Systems Corporation	Automated bloodbank management
Advanced Micro Devices, Inc.	Semiconductor manufacturing
Alza Corporation	Ethical drug research
Angerics, Inc.	Health care
Automatix, Inc.	Robotics
The Children Place, Inc.	Specialty retailing
Company Stores Development Corporation	Manufacturers' outlet malls
Data 100 Corporation	Computer peripherals manufacturing
Delphi Communications Corporation	Automated telephone answering service
Directions Systems, Inc.	Outdoor advertising
ECS Microsystems, Inc.	Distributed processing systems manufacturing
Education Today Company, Inc.	Magazine publishing
Energetics, Inc.	Energy systems manufacturing
Four Phase Systems, Inc.	Distributed data processing
Garrett Music Enterprises	Records production
Geosource, Inc.	Oil services and products
Index Systems, Inc.	Computer software development
Instapak Corporation	Specialty packaging
Mark Controls Corporation	Value manufacturing
Media Networks, Inc.	Magazine advertising
Menlo Trading Company	Specialty retailing
Microform Data Systems, Inc.	Microfiche systems manufacturing
Modular Computer Systems, Inc.	Minicomputer manufacturing
Molecular Genetics, Inc.	Microbiological research
Paradyne Corporation	Data communications equipment manufacturing
Quanah Petroleum, Inc.	Oil and gas exploration
Ropart, Inc.	Women's active sportswear retailing
Royce Electronics Corporation	Citizens band radio marketing
S.A.Y. Industries, Inc.	Specialty packaging
Sealed Air Corporation	Specialty packaging
SEI Corporation	Bank trust data processing services
Shugart Associates	Flexible-disc drive manufacturing
Spaid Industries, Inc.	Fluid power products
Staid, Inc.	Point-of-sale systems manufacturing
Systemedics, Inc.	Medical data processing
Tesdata Systems Corp.	Computer efficiency measurement
Thetford Corporation	Chemical waste handling
TRU, Inc.	Automotive tune-up centers
VLI Corporation	Health care
Weeks Petroleum, Ltd.	Oil and gas exploration

Table XX.   Management Buy-out Companies

Name of Company	Industry
APCOA, Inc.	Parking lot operations
Archon, Inc.	Health foods manufacturing
Baldt, Inc.	Marine hardware manufacturing
Belvedere Holdings Ltd.	Reinsurance
Beverage Distributors Corporation	Beverage distribution
Cosco Industries, Inc.	Marketing service, manufacturing
Cott Holding Corporation.	Soft drink bottling
Drake, Smith Company, Inc.	Furniture manufacturing
Dreyfus Prime, Inc.	Marine hardware distribution
Elsco Industries, Inc.	Lighting fixtures manufacturing
Everfast, Inc.	Fabric retailing
Illinois Coil and Spring Company	Spring and controls manufacturing
Industrial Distributors of America, Inc.	Industrial supplies distribution
James River Corporation	Paper manufacturing
Lindon Chemicals & Plastics, Inc.	Chlorine and caustic soda production
Mid-Continent Industries, Inc.	Soft drink bottling
Presco Holding Corporation	Nonstandard automobile insurance
Rob Ray, Inc.	Boys' sportswear manufacturing
Sinclair & Rush, Inc.	Dry-molded plastic component manufacturing
Unimet Corporation	Metals distribution
Vdk Holding Corporation	Bakery products manufacturing
Vitreon, Inc.	Glassware manufacturing
Washington Chain & Supply, Inc.	Marine hardware distribution

three categories: (1) Emerging Growth Companies, or new product investments; (2) Management Buy-out Companies, or firms divested from conglomerates; and (3) Turnaround Companies, or firms purchased and restructured. This list provides an overview of the variety of products and services targeted for capital venture funds. The company names marked by asterisks are currently in the Sprout Group's portfolio.

Evaluation of the Management Technique

The Sprout Group will continue to grow. A recent acquisition is a company called Molecular Genetics, which produces enzymes by genetically altering bacteria. Venture capital investment will remain on the rise

Table XXI.   Turnaround Companies

Name of Company	Industry
Baxter, Kelly & Forest, Inc.	Velour manufacturing
Clopay Corporation	Housewares manufacturing
Envirotech Corporation	Mining and pollution control equipment
Fox-Vliet Holding Corporation	Drug wholesaling
New Puro Associates	Water cooler rental
Polaris Resources, Inc.	Mining
RSR Corporation	Lead recycling

for several reasons. First, new federal tax laws reward individuals, corporations and institutions that support entrepreneural development or conglomerate divestment. Second, new issues of stocks have fared well, drawing investors out to purchase them, which creates interest in alternative investment opportunities. Moreover, an increase in pension funds, other than tax-exempt funds, is supplementing money available to venture capital markets.

The Sprout Group has not been a member of what has been a very "clubby" industry. Instead, it has attempted to maintain strong ties with the parent firm, Donaldson, Lufkin & Jenrette, Inc., for management review support. In future, Sprout plans to work more directly with other venture capitalists, who are well-positioned geographically or have some special industry expertise. Attempts also will be made to initiate more investment opportunities by combining technical or product expertise with business managers capable of capitalizing on product opportunities.

Other organizations and institutions employing the technique are the following:

- University of Wisconsin
- National Aeronautics and Space Administration
- Argonne National Laboratories

FRANCHISING—McDONALD'S CORPORATION

Introduction

Background

Franchising is a term used to describe the legal relationship between two business entities. A franchisee purchases from the franchisor a given set of licenses and agrees to execute the business according to the principles of the franchisor. The franchisee's operation generally is separated physically from the franchisor, yet the franchisee may receive all or some of its products from the franchisor. Usually, the franchise agreement requires that the franchisee give the profits of the franchise operation to the franchisor. The franchisee, in turn, receives a salary based on a percentage of his operation's total sales. The franchisor often provides management training, operational equipment, facility design and national marketing services.

This management technique is effective in the technical innovation process because it assists in the diffusion of new technologies. Also it provides an environment that stimulates new product development. Franchising is a management tool that has been used in a number of industries such as small computers, hotels and motels and the fast-food service. To exemplify the application of franchising as a management tool to encourage the diffusion of technologies, McDonald's Corporation and its operations are discussed below.

Organization

The McDonald's Corporation is a franchise operation of fast-food restaurants. It boasts of being the largest food-service organization in the world, with more than 6000 restaurants located in all 50 states, as well as 26 countries outside the United States. Gross revenues in 1980 totaled $6.2 billion, up 16% over 1979. The corporation began in 1955 when the founder purchased the first store. His intent was to open other stores in which he could install multimixers (multiple malted milk mixers) that had been invented by a friend. This giant franchise operation began more than 25 years ago with a desire to promote technology transfer.

Franchises, including McDonald's, are not structured to encourage innovation by the franchisee. Rather, the development of new technological products and process designs occurs at corporate headquarters. There the food service operation is tested and reviewed continually to assess the need for changes. Often, new equipment ideas will be researched by the corporate engineering staff and developed by small manufacturing firms to which licenses for the new products have been granted.

The McDonald's franchise operation is distinctive as it is a highly automated food product service system and has high quality standards. Equipment and food are supplied almost exclusively through the franchisor/headquarters. The facilities are built by the corporation using a common design framework to support the highly automated services.

There is literally no room within the operating facilities for experimentation with, or use of, a new technology. Although some franchisees within the McDonald's Corporation have attempted to introduce new food items to the menu, the corporation is more encouraging of innovation at the top level of the organization.

Management Technique

Overview

The franchisor-franchisee relationship is based on mutual financial/management needs. It is highly dynamic and ever changing. Both parties have responsibilities to each other, as outlined by the franchise agreement. Part of this agreement ensures that the franchisee will adopt new technology as recommended by the corporate headquarters to maintain a peak level of automated operation. Technologies developed by McDonald's equipment engineers and built by independent manufacturers include computer programs, which control a number of the system's cooking procedures, polystyrene packaging for the hamburgers, a hand-held device to dispense sauce, an audio-visual timer system used in cooking and a bun toaster. Direction, control and financial support of the innovation process is provided by the corporate management.

Selection and Implementation of
the Management Technique

One of the keys to McDonald's successful adoption of technological innovation through franchising is the involvement of local business persons in management of franchise operations. A franchise operated by an established member of the community grows rapidly. As a result of the successful establishment of the franchise operation, the local franchisee usually welcomes new technologies or management approaches provided by corporate franchise headquarters.

Evaluation of the Management Technique

Evaluation of franchising to determine its impact on the innovation process can be accomplished by considering each phase of the process and measuring actual performance of the organization against anticipated performance or competitors' performance. To achieve the goal of assuring use of common equipment, McDonald's franchisees must accept all new technologies used to provide food preparation service.

No other organizations employ this management technique.

LICENSING—BELL LABORATORIES

Introduction

Background

The technique of licensing new products and processes links the inventor/developer with the manufacturer/marketer. It is a valuable mechanism for transferring technology and has been used to overcome barriers that occur primarily in the final phase of the innovation process, affecting diffusion of new technological developments. Bell Laboratories, a research institution and subsidiary of the American Telephone and Telegraph Company (AT&T), has utilized licensing techniques to stimulate interest in the semiconductor business and to increase access to state-of-the-art technology.

Organization

A fairly detailed description of the Bell Laboratories organization is presented in the discussion of personnel transfers. As noted there, Bell Laboratories is a part of a highly technical industry that sustains a strongly competitive environment. As a result, both Bell and Western Electric, AT&T's manufacturing subsidiary, use (1) patents to protect new products and processes; and (2) negotiate licensing arrangements to control the duplication and diffusion of the new products and processes developed within AT&T. The licensing procedures initiated by AT&T include cross-licensing privileges, which enable AT&T to have access to the licensees' patents. The amount of money collected by AT&T in royalties on licenses is dependent on other types of benefits gained through the licensing arrangement. Benefits may permit AT&T to access a licensee's patent portfolio and be privy to certain R&D activities. AT&T usually reassesses its licensing agreements every five years to determine their validity and appropriateness and the associated benefits relative to AT&T's research needs. Licensing arrangements that have been instituted by AT&T have increased the rate of certain technological developments within the industry.

Management Technique

Overview

The licensing of technology is an attractive device for marketing new products or processes because it delays setting a price for the new concept.

Pricing can be difficult as additional development may be required or the products/processes may need to be tested in the marketplace. If a new product's potential market value could be estimated, it could be sold directly rather than allowing both the licensing agent and licensee to profit from sale of the product.

Licensing of new products may lead to the following situations:

- When a new product has been developed by a firm with limited marketing capabilities, the value of the technology can be increased if licenses to manufacture and sell the product are granted to other firms.
- When a new product is developed by a firm, but the product does not lie within the company's principal business, a spin-off technology can be more easily marketed by a firm in the relevant industry.
- When the company that has developed the new product does not have the financial resources, it cannot support adequate commercial development of the new product.
- When two companies want each other's proprietary technology, the rights simply can be traded through a cross-license.
- When a company has developed proprietary methods for manufacturing an item important to its own needs, but it cannot be bothered with the actual manufacture.

An important consideration as to whether a new product or process should be licensed and what the general licensing strategy of a firm should be also depends on whether a market exists for the new product or whether one must be created.

*Selection and Implementation of
the Management Technique*

The prime reason firms wish to obtain licenses permitting use and development rights for new products is that they do not have the technical or financial resources to carry on research in the new product area. On the other hand, the firms choosing to obtain licenses have expertise in the areas of marketing and manufacturing, as well as financial resources to invest in the development of the new product. There are firms that act as facilitators to locate prospective licensees. Expositions and trade shows also emphasize licensing opportunities.

The most valuable prospective licensee is a firm that has considerable entry into the product area but does not have a substantial psychological or financial investment in a preexisting development program. Another good prospective licensee is a company that is already active in a particular field but is at a competitive disadvantage or is operating under another license with which it is dissatisfied.

Bell Laboratories has utilized the forum of symposia to educate its licensees regarding new technology developments. In the area of semi-conductors, Bell initiated its series of symposia in 1951 with lectures on the transistor. In 1952 Bell sponsored an eight-day symposium to describe transistor properties and applications, as well as the related physics and technology: 25 American firms and 10 foreign firms, licensees of AT&T, paid $25,000 in advance royalties to participate. Another symposium for licensees was conducted in 1956 to review technological advances such as diffusion and oxide-masking. Since this 1956 symposium, Bell has relied on individual visits by licensees, rather than symposia, to pass on information regarding technological developments, such as epitaxical techniques in the semiconductor industry.

In offering its licensees valuable knowledge concerning new tech-nology, Bell Laboratories' main objective is to establish a formal com-munications sharing network between the labs and the licensees, thus discouraging potential licensees from "stealing" ideas without sharing technically or financially in the development of new products and processes. It is difficult to keep new developments secret when the scientists and engineers are highly mobile. In the semiconductor industry, new small firms have been particularly reticent to infringe on patent rights. But by offering attractive licensing arrangements, AT&T has been able to convince potential rivals to accept its patents.

Evaluation of the Management Technique

Licensing has proven to be a valuable intraindustry tool to encourage innovation and yet maintain competitive interests among industry members. Its successful application is dependent on the technology to be licensed, the reliability of the patent system relative to that technology and the ability of the license participants to evaluate accurately the worth of the license and, thereby, to assess appropriate benefits to the participants.

The high costs, high risks and uncertain time scale of internal R&D programs have led to genuine economic incentives favoring acquisition of outside technology. Particularly for organizations lacking a manufactur-ing and/or marketing capabiltiy, such as individual inventors, universities and private research organizations, licensing is an effective means for commercializing new inventions. Licensing appears to provide more controlled diffusion of a new technology than the use of patents. Patents are intended to protect an inventor from having the same invention made and sold by other individuals. When a patent is granted, the new technology information becomes public without control over use or copying. If a patent is licensed, however, its strength is measured by its ability to be recognized and respected by potential infringers and licensees, preferably without litigation.

Other organizations employing the technique are the following:

- International Harvester Company
- Argonne National Laboratories

MIDDLEMAN—THE MITRE CORPORATION

Introduction

Background

During the 1970s, the University-Industry Cooperative Research Centers Experiment was conducted by the National Science Foundation (NSF) to test and evaluate management approaches used to stimulate nonfederal investments in R&D and to improve industry's use of its results. The basic concept was to link industries with universities to increase the rate of industrial innovation.

During the first year, 14 organizations were funded to obtain commitments from targeted industries. At the end of this year, three organizations were selected for four-year funding: (1) the Furniture R&D Applications Institute at North Carolina State University; (2) the MIT-Industry Polymer Processing Program at the Massachusetts Institute of Technology; and (3) the New England Energy Development Systems (NEEDS) Center within The MITRE Corporation.

This management technique encompasses the use of a middleman—in this case The MITRE Corporation—by the industry and universities to recruit cost-sharing participants in the project and coordinate R&D activities conducted by the universities (Figure 9). This experiment was designed to affect technical and financial barriers that inhibit industry's access to technical expertise. The middleman acts as the key link among the university research members, the industry cost-sharing members, the university data collection and analysis team and the funding/experimental evaluation conducted by NSF. The experiment was completed in 1979.

Organization

During formation of the NEEDS Center, it was expected that the Center's activities would be coordinated by a not-for-profit center

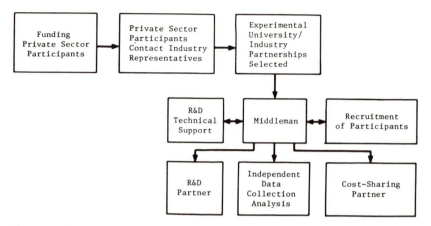

Figure 9.   Role of the "middleman"—a cost-sharing partner to industry and an R&D partner to universities.

separate from The MITRE Corporation. However, it became apparent that the center's activities could benefit from direct access to, and association with, MITRE's technical resources. The MITRE Corporation is a not-for-profit research and engineering firm that was established in 1958 primarily to provide technical and communications consulting services to the U.S. Air Force. This charge has expanded during the past 20 years, and although more than half of the work performed by MITRE is for the Air Force, slightly less than half of MITRE's work is conducted in a variety of scientific and technical disciplines. A partial listing of MITRE's programs and major sponsors is presented in Table XXII.

In 1980 MITRE's revenues totaled $153 million. MITRE has two large sites, located at Bedford, Massachusetts and McLean, Virginia, as well as approximately 20 domestic and foreign site offices. Technical personnel totaled more than 2000 at the end of 1980. The NEEDS Center was established within the firm in 1974. Figure 10 presents the relationship of the NEEDS Center within the corporation and Figure 11 indicates the general relationships of the participants in the NEEDS Center program. A discussion of the participants' roles is provided in more detail in the next section.

Management Technique

Overview

The NEEDS Center served as a middleman to link 20 universities and schools with 28 organizations in the private sector for cooperative

Table XXII. MITRE's Program Offices and Major Sponsors

Technical Work Areas	Sponsors
Air Defense Tactical Air Command and Control Strategic Command and Control National Level Command and Control Battlefield Systems Surveillance and Target Acquisition Military Communications	Department of Defense
Energy, Resources and the Environment Computers and Software Civil Information Systems Components and Technique Development Air Traffic Control	Department of Agriculture Department of Commerce Department of Energy Department of Health and Human Services Department of the Interior Department of Justice Department of Labor Department of State Department of Transportation Department of Treasury Environmental Protection Agency

research. In this coordinating role, the NEEDS Center had four major functions:

1. to engage in R&D, primarily as a middleman, serving industry as the cost-sharing partner and universities as the R&D partner. MITRE also participated in the NEEDS Center Program as an R&D performing partner;
2. to disseminate results of NEEDS Center projects throughout the energy industry.
3. to recruit participants from industry and R&D programs at universities to become involved in the NEEDS Center program activities; and
4. to collect and analyze data concerning the projects undertaken.

The seven utilities servicing the northeast sector of the country provided cost-sharing funds. As a group, the utilities donate approximately 1% of sales to research, of which an average of 22% was conducted in-house. The 12 private organizations that financed the NEEDS Center collectively spent approximately 2% of sales to support research. This group consisted of large corporations with at least a portion of their business in the energy field. A third group, made up of cost-sharing participants in the NEEDS Center Program, consisted of nine not-for-profit organizations that shared a public interest in the center's activities.

Data collection and analysis in the NSF experiment were conducted by

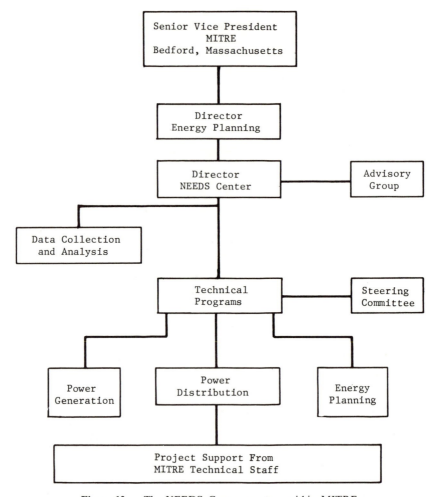

Figure 10.   The NEEDS Center structure within MITRE.

the University of Texas to determine whether the NEEDS Center increased investment in R&D by utilities and to identify the characteristics of utilities that could best explain their different attitudes toward investment in R&D.

Approximately $1.4 million in cash and in-kind contributions were provided by the private sector to support 18 projects at the Center. These projects each addressed one of the following applications:

- Development of hardware to be integrated into an existing system
- Development of a technique to assist in solving an operational or informational problem to be incorporated into an existing process

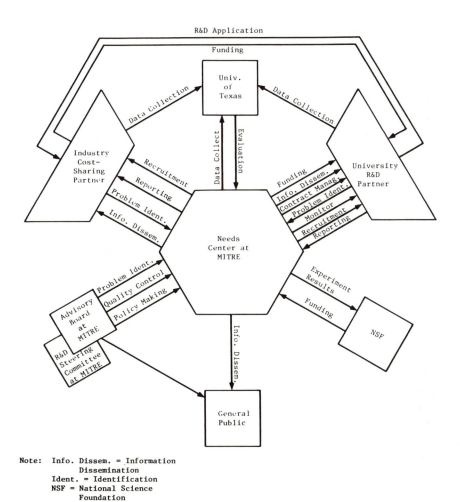

Note: Info. Dissem. = Information
              Dissemination
      Ident. = Identification
      NSF = National Science
            Foundation

Figure 11. Relationship of participants in the NEEDS Center program.

- Design for service
- A study to define problems or evaluate alternatives as a basis for future R&D
- Conducting a seminar or workshop

These projects followed part of or all seven stages of research—recognition of the problem, problem definition, research and development, solution, demonstration, evaluation and application. Three projects followed through all seven stages. Eight projects began at problem definition and showed medium progress toward reaching the final stages. The other projects were initiated during the final stages of the R&D process, requiring much less time to complete and less support from the Center.

Selection and Implementation of
the Management Technique

The selection of this "middleman" management technique during the experimental process evolved as the need was recognized for a not-for-profit organization with sufficient technical expertise to play the coordinating role in the project.

The MITRE Corporation provided administrative, technical and financial support for the Center's research activities. To illustrate the role of MITRE as the middleman, a description of the activities related to one of the projects at the Center is presented here. The project was entitled "Power Plant Dynamics, Simulation, and Control." Its goal was to develop mathematical models of two power-generating units and control system design specifications. The R&D performers were the staff at the NEEDS Center and M.I.T. Funding for this project was provided by The MITRE Corporation, the Boston Edison Company (BECO) and NSF. This project covered a four-year time period, from 1972 to 1976.

It actually consisted of four small projects; two were independent R&D programs that together effected the solution to a specific power generation problem in an electric utility. The other two related to communications, or how to publicly disseminate information about the Center's work in boiler modeling and to facilitate the exchange of information and the cross-fertilization of ideas between the developers and users of this analytical tool.

In 1972 representatives of The MITRE Corporation and BECO met to explore the possibility of obtaining MITRE's assistance in solving some of BECO's problems. Seeking alternative solutions, BECO already had contacted equipment vendors and industrial consultants but had not made any progress. The company turned to MITRE in hopes that its systems engineering capabilities could offer a new problem-solving approach. MITRE sent out a team to evaluate the scope of the problem and performed an analysis of the industry as related to the stated problems. MITRE then applied to NSF to participate in the Industrial Innovation Program.

After the NEEDS Center was established in 1973, BECO's problems became the keynote project. Since the project problems had been defined clearly, the next step was to recruit utility support and to develop the mathematical models. During 1974–75 MITRE acted as a cost-sharing partner as well as the research and development partner, while the

NEEDS Center staff visited potential project supporters. Once the cost-sharing arrangements had been made, M.I.T. was invited to participate in model development.

To increase MITRE's credibility as a middleman serving industry, a seminar on boiler modeling was held. The NEEDS Center invited a panel of experts on related topics to present papers dealing with the development, application and utilization of boiler modeling. The NEEDS Center's management responsibilities were to disseminate information and conduct the seminar and project.

Evaluation of the Management Technique

The relative success of the NSF experiment was evaluated in terms of: (1) the NEEDS Center as a middleman; (2) universities as R&D performers; and (3) private sector organizations as cost-sharing partners. Based on the measurable financial support obtained by the NEEDS Center, its success was attributed to the following characteristics of the "middleman" organization:

- Ability to enlist support of MITRE's top management
- Ability to enlist the support of opinion leaders in the region
- Concentrated effort and persistence in launching projects
- Established credibility with potential project participants
- Ability to identify problems requiring immediate solutions

In the Power Plant Dynamics project, one of the first projects started at the Center and industry-university linkages were strengthened by the interdisciplinary conference carried out within the R&D program. Although the development of flexible models, the result of the project, provided a solution to a long-standing problem, the anticipated continued support for these modeling projects by industry was not realized fully.

PERSONNEL TRANSFERS—BELL LABORATORIES

Introduction

Background

Personnel transfers—or the movement of personnel within an organization or between organizations in industry—allow individuals to pursue more stimulating career opportunities and transfer experience or knowledge to another part of the firm. Management may respond positively or negatively to personnel transfers initiated by employees. The management technique described in the following sections involves personnel transfers encouraged by management.

The telecommunications industry offers perhaps one of the best examples of the occurrence and positive reinforcement of personnel transfers. Bell Laboratories, the research arm of AT&T, has been, during the past 30 years, a prime motivating agent for development of new products and processes as a result of personnel spin-offs from the labs to form new firms.

The major barrier affected by this management technique is the inaccessibility or unavailability of technical information required to produce technological innovation. Particularly in industries in which discoveries are commonplace, as in the telecommunications industry, the normal channels through which technical information passes, (e.g., journals, professional writings) cannot facilitate information transfer as quickly as word of mouth or transfer of information by relocation of personnel. This transfer process may occur at any stage in the innovation process and may result in the formation of new products and processes or adaptation of existing ones.

Organization

Bell Laboratories was established more than 50 years ago to develop new telecommunications-related products and processes. These are fed to Western Electric, the manufacturing arm of AT&T. The new products and processes are produced by Western Electric and distributed to the operating companies within the nationwide system.

Bell Laboratories (labs) employs more than 17,000 persons, including approximately 6000 scientists and engineers. In addition to four major centers, the labs include eight other laboratories and a number of field centers. The organizational framework at the labs follows categorical lines—research, systems engineering and development; 90% of the work at the labs is in systems engineering and development and approximately 10% is in research. Half the research budget is earmarked for basic research and half for fundamental communications systems research.

The structure of the organization is such that there is sufficient elasticity and overlap of functions to promote interaction among the members of the technical staff. The professional community at Bell Labs maintains a conscious awareness of the need to increase interfaces among the staff to support a creative innovation climate. The interaction is encouraged by management through geographic deployment, recruiting policies and exchange of personnel among functional divisions.

Management Technique

Overview

Personnel transfers from a firm would appear to be counterproductive in terms of losing scientific talent valuable to the firm. Most firms that actively support these transfers view it as a positive influence on the business as this posture not only will attract top talent, but strengthen the communications flow within the innovation community, which is important to maintaining a high degree of creativity.

There are two broad categories of personnel transfers: (1) those initiated by management to rapidly transfer experience and knowledge from one area of the firm to another; and (2) those initiated by individuals, most notably from one firm to another. Both categories affect rapid diffusion of technology and the development of new products and processes. The discussion that follows focuses on interfirm personnel transfers.

Selection and Implementation of
the Management Technique

Bell Laboratories is considered one of the major pioneers in the development of telecommunications. Since the corporation was formed, more than 50,000 patents have been issued to the firm, of which approximately 2000 are currently in force. Some of the outstanding inventions are bubble memory, coaxial cables, stereo sound recording, negative feedback and the transistor. The innovation climate at Bell Labs has been attractive throughout its history because it has maintained strong links with universities to recruit and educate staff as well as to exchange scientific research. It has maintained an atmosphere conducive to professional recognition and reward. There have been other factors that have drawn personnel away from the labs, and it has been the philosophy of management to acknowledge and, in some cases, support personnel transfers.

In the 1950s, Bell Laboratories employed most of the scientists and engineers in the semiconductor industry. Bell encouraged the development of solid state reseach within the industry through publications of scientific progress in the field. It became apparent that smaller firms were more adept in taking technology from the laboratory and adapting it for large-scale production. Although these small firms rarely captured a major market share, their presence quickly promoted diffusion of new technology. Figure 12 illustrates a small set of firms that spun off from

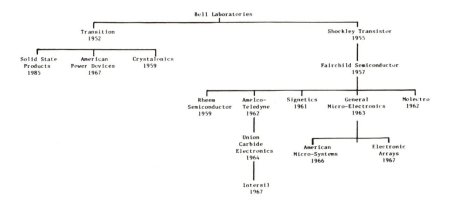

Figure 12.   Spin-off firms of Bell Laboratories.

Bell Labs in the 1950s and ultimately grew into a generation of new firms developing new technology. Individuals were able to spin off from the labs due primarily to an easy access to technology and the availability of venture capital, coupled with insignificant scale economies and government demand and support.

Like other businesses in highly technical and competitive industries, Bell Laboratories patents any new development that is patentable, and Western Electric negotiates licensing arrangements between AT&T and other firms, often with a cross-licensing provision granting AT&T access to the licensee's patents. Royalty payments depend on the patent portfolio, as well as on research and development capabilities of the licensee. Licensing agreements usually are reassessed every five years. By offering licensing arrangements to other firms, they have accepted their patents in some cases, which opened access to important technological developments within the industry.

One firm that attended a license conference sponsored by Bell Labs was a small instruments and geophysical exploration firm, Texas Instruments, which took a license from Bell and privately embarked on a major program of research, development and investment. Within 15 years its sales jumped from $20 million to $3 billion. Because of the Bell Labs' liberal licensing policies and encouragement of interfirm mobility for scientists and engineers, new firms were formed rapidly in the semiconductor industry.

Evaluation of the Management Technique

Personnel transfers from Bell Laboratories have occurred during recent years as a result of the state-of-the-art of the development of technology, the position of the Labs as a research entity and the role of AT&T as a major technology innovator. A number of factors have changed the structure of the industry, as well as the role to be played by the Labs as the continuing pioneer in telecommunications. In the 1950s, when Bell Labs started to spin off a number of new firms, its image was that of a closed, paternalistic research organization. In recent years, work at the Labs has become much more market-driven. Individuals used to leave Bell Labs because professional staff members harbored conflicting views of the Labs' purpose. Some believed the Labs should act as a product design and development arm of AT&T; others believed it should be an academic, pure research organization.

Generally it has been accepted that mobility of personnel, either within or between organizations, balances the flow of technical communications. Although there appears to be no evidence to quantify this phenomenon, comparison of American technology firms which encourage personnel transfers, with European or Japanese technology firms which reward firm loyalty (stable employment), indicates that the rate of technology transfer is notably faster in the American setting.

Another organization that employs this technique is the National Aeronautics and Space Administration (NASA).

UNIVERSITY-INDUSTRY LINKAGES— UNIVERSITY OF WISCONSIN SMALL BUSINESS DEVELOPMENT CENTER

Introduction

Background

The University of Wisconsin Innovation Service Center was established in 1980 under the direction of the University of Wisconsin Small Business Development Center. Its purpose is to provide opportunities for education and research for the innovator, as well as to supply public services related to coordination of the university with public agencies and industry. The major barriers are the lack of technical expertise available to the entrepreneur during the research and development stage of the innovation process and the inability to analyze the marketability of the new products or processes.

The Small Business Development Center network was set up by the U.S. Department of Commerce to serve as a link between management and technical support services and small business operators. Several national innovation conferences held in the late 1970s revealed a critical need to educate innovators and entrepreneurs regarding the business aspects of developing and marketing new products and processes. It was pointed out that the number of small businesses introducing innovative products had declined markedly due largely to competition from big business and the innovator's inability to overcome process barriers.

The Small Business Development Center network was patterned after the successful County Agricultural Extension Agent program established by the U.S. Department of Agriculture (USDA). The county Agriculture Agent program places a USDA representative in each U.S. farming county. The agent's purpose is to work with the farmers, who also are small business managers, providing them with the technical expertise required to maintain an awareness of the state-of-the-art of technology, as well as knowledge regarding financial management of farming operations. The Small Business Development Center program is different in that its centers often are based at universities at the state, rather than the county, level.

The Small Business Development Center at the University of Wisconsin was established in 1978. It serves the entire state with a population of approximately 5 million people and an area of 56,154 square miles. Wisconsin leads the nation in milk and cheese production. In addition to farm products, the chief industrial products are automobiles, machinery, furniture, paper, beer and processed food.

Organization

The Center includes 11 service centers located at university extension sites around the state. An administrative staff consisting of a director, an associate director, a grant writer/editor and three program assistants coordinates the following three major program activities: (1) business outreach programs; (2) a small business institute; and (3) workshops and conferences. Each service center is directed by university faculty and graduate students participating in the small business institute. The Wisconsin Innovation Service Center acts as an integral part of the Small Business Development Center.

Management Technique

Overview

As mentioned above, there are three program functions served by the Small Business Development Center. The business outreach program is one of the major thrusts of the Small Business Development Center's activities. This program offers free information services to small business operators who contact the centers requesting specific information. Typically, an encounter between a small business owner and representative of the Center will require less than 12 hours of assistance. Depending on the requirements of the business problem, the requesting individual may be referred to training and/or educational resources provided by the Center. The requesting individual also may be referred to a faculty or staff member who has special expertise directly related to the problem. Another option is to have the requesting individual referred to a private consultant to attain paid counseling or services beyond the scope of the training and educational services offered at the Center.

The small business institute was established to provide business school graduate students with opportunities to conduct case studies of business problems. Small business operators who have approached the Center are screened by its administrators to select those business-related problems that lend themselves to study over a four- to five-month time frame and offer a unique study objective. Graduate students are assigned a case for a semester. The student works with the small business operator to characterize the problem, formulate alternative solutions and recommend a plan of action. The recommended plan, prior to implementation, is reviewed by the small business operator, other students participating in the case study exercise and the sponsoring professor. The small business operator also can evaluate the student's performance at the close of the project.

The workshops and conferences sponsored by the Center generally focus on business problems identified during the counseling sessions or small business institute case studies. An example of one of the workshops is the Invention Workshop, which is directed by a former business executive who worked in the field of robots and automated industry and has invented several new products and processes in that field. Discussions during the Inventor's Workshop focus on issues such as cost-effective decision-making, patent procedures, technical assistance available to independent and small business inventors and management assistance with marketing, financing, prototyping, patent law and patent searches. Other speakers participating in the workshop are professionals from banking institutions, the Patent Law Association, private business and industry, universities and government, as well as executives in sales and marketing.

Selection and Implementation of
the Management Technique

The University of Wisconsin Innovation Service Center was established within the Small Business Development Center primarily to bridge the gap between the innovator or entrepreneur, and industry, and to assist the small businessperson in overcoming product development barriers. Some of the factors influencing the development of new products and processes outside large industry are increasing costs and market complexity. Costs for evaluation, business analysis, development and market analysis are often high. Frequently, considerable technical and financial resources are required to develop an idea to meet federal, local and industrial standards and regulations. Often it is difficult (and expensive) for the innovator to ascertain whether the new product is commercially viable. Moreover, an innovator may find it difficult to get a new product or process accepted by industry due to the individual corporate and market feasibility demands, such as manufacturing facilities, existing product lines, well-defined markets and financial return requirements. The Innovation Service Center aims at easing these problems by maintaining relationships with state government representatives to try to improve the environment for noncorporate inventors and innovators.

A valuable service provided by the Innovation Service Center is the evaluation of new products and processes using a formal evaluation method to determine the development potential of an idea or invention. The preliminary innovation evaluation system attempts to answer the following three questions:

1. "Will it work?" (technical feasibility)
2. "Can it be sold?" (commercial feasibility)
3. "What type of corporation can sell it?" (corporate feasibility)

One attractive feature of the evaluation procedure is that it standardizes evaluation criteria, making communication between the entrepreneur and evaluator easier, more efficient and more effective. Commercial feasibility, however, is determined in the marketplace rather than by the opinion of an evaluator, even if it is based on a set of structured characteristics or circumstances. New products result primarily through a deliberate campaign in which someone has championed the product.

Evaluation of the Management Technique

The goal of the Small Business Development Center at the University of Wisconsin is to provide an intensive statewide delivery network of programs to assist businesses in achieving profitable growth. Specific program objectives are:

1. to become a principal factor in the development of the economy of the state of Wisconsin;
2. to provide relevant and meaningful education and training designed to fully utilize the tremendous resources of the University; and
3. to perform substantial and continuous basic and applied research oriented to business management that will be carried out by faculty, graduate students and the Center's full-time professional staff.

The best measures of the effectiveness of the Center is provided by the small business operators. Responses to the short questionnaire issued to all Center users indicates overwhelming support of the services. The future of the Center will depend on continued funding by the federal government, the ability and interest of the state and University to assume more of the Center's financial responsibilities, and possibly the willingness of Center users to pay for services rendered.

Another organization that uses this technique is NASA.

INDUSTRIAL APPLICATIONS CENTERS— NATIONAL AERONAUTICS AND SPACE ADMINISTRATION

Introduction

Background

A wide range of technologies designed for a multitude of purposes has been developed as a result of NASA's programs during the past 30 years. The initial applications of these technologies generally were limited to scientific fields within the aerospace industry. In an effort to increase the use of the technological developments resulting from the space programs, NASA established its technology transfer program. This program promotes secondary applications of aerospace technology by (1) disseminating information on technologies available for transfer; (2) assisting industry in the transfer process; and (3) adapting existing aerospace technology to solving problems in the public sector. The program has accomplished this objective through problem identification, direct technology assistance and cooperative projects. Training, demonstration projects and grant funds are used to assist in the transfer process. Technical assistance and information dissemination centers also have proven valuable in assisting both public and private sectors in the development of new technology applications.

This section describes NASA's Industrial/Regional Applications Center network as a mechanism for overcoming barriers in the innovation process associated with an industry's lack of technical expertise and an inability to support expensive R&D.

Organization

NASA has established five Industrial/Regional Applications Centers that serve the public and private sectors, assisting users in identifying technical information that can alleviate product- or process-related problems. The primary information data base includes more than 800,000 scientific and technical aerospace reports. Linked with this data base are abstract services covering the fields of chemistry, engineering, electronics, plastics and metallurgy.

In concert with the regional dissemination centers, NASA operates seven industrial applications centers and two State Technology Applications Centers. These centers use NASA research and engineering teams representing different areas of expertise to assist in identifying public and industrial problems that may be subject to technological solution. The role of these teams is to (1) identify existing NASA technology that may be applied to the problem areas; (2) identify markets for products incorporating the new technology, and (3) disseminate the information. Projects addressed by the technical applications teams may result from requests for NASA assistance from other government agencies or from NASA technologists recognizing other applications of the NASA-developed technologies. The technical applications teams, along with the regional dissemination centers, are located at research institutes and universities. They concentrate on providing assistance in the fields of health care, public safety, transportation and industrial productivity.

NASA contractors responsible for developing new products and processes are responsible for preparing written reports describing their inventions, innovations or improvements. These reports are included in NASA's technology utilization publication: *Tech Briefs*, distributed quarterly to more than 60,000 industrial subscribers. Approximately 150 new products/processes are discussed in each issue. If individuals are interested in receiving additional information concerning a specific product or process after having reviewed *Tech Briefs*, a technical support package can be requested from NASA that provides more detailed information regarding the innovation. Approximately 200,000 requests for technical support package materials are received every year.

Management Technique

Overview

The regionally located information and technical support services (dissemination centers, industrial applications centers and technical assistance teams) provide services in defined geographic areas. The core of technical expertise is drawn from the NASA staff, and the NASA technology transfer network receives assistance from universities, as well as other research and engineering concerns. The users of this network have access to a wide variety of resources, including more than 10 million technical documents covering every field of aerospace-related activity and the contents of 15,000 scientific and technical journals.

The industrial applications centers actually provide the following three basic types of sources:

1. retrospective searches of the data base to identify abstracts or reports on subjects related to the company's needs;
2. "current awareness" services—tailored, periodic reports designed to keep a firm's executives aware of the latest developments in a specific technology area; and
3. technical assistance to apply the information retrieved from the NASA data base.

A nominal fee is charged by the centers for their services.

Selection and Implementation of the Management Technique

One NASA technical support team has been instrumental in the development of a metrorail rapid transit system in the Miami, Florida area. NASA became involved in the project as a result of discussions between the staff at NASA and the Office of Transportation in Dade County, Florida, the city of Miami and the Kennedy Space Center regarding applicability of NASA technology to the project. As a result, the technology applications team from California started a program to apply NASA engineering and management technology to Metrorail problem areas. A NASA-trained engineer became a full-time consultant with the Office of Transportation to examine transit design problems, to identify areas in which NASA already had achieved applicable solutions and to serve as coordinator and liaison with other NASA technical resources. Technical information concerning risk and configuration control and hardware technologies such as anticorrosion measures, fire and lightning protection, solar energy utilization and materials selection were transferred as a result of the consultant activities.

Evaluation of the Management Technique

A review of a set of technologies that has been transferred successfully from NASA to the private or public sector has been conducted [6]. This study showed that the success of the technology transfer is directly dependent on a number of factors:

1. the relevancy of the technology to the firm's business area;
2. strength of the relationship between the technology and the problem to be addressed;
3. urgency with which the problem needs to be treated;
4. quality of information regarding the description and use of the technology;
5. level of maturity of the technology;
6. availability of personnel required to implement the technology;
7. availability of financial resources necessary for implementation; and
8. top management's interest in adapting the technology.

This research also indicated that involvement of the NASA innovator in the transfer process appeared to be supportive of the successful adaptation. Also, the presence of a product champion, or an individual who went beyond his organizational role to bridge the organizational and intergroup gaps to promote the product idea, tended to be an important positive factor in the adoption of the innovation.

More than 60% of the U.S. personnel trained in research, development and engineering work on large government-related projects. It is absolutely vital that the fruits of this talent be channeled to the private sector through formal technology transfer mechanisms. The aerospace-related contributions in such fields as health care, computers and remote sensing have been outstanding. Although the transfer of these technologies might have occurred naturally, the NASA technology transfer program has improved and accelerated the process considerably.

No other organizations employ this technique.

INNOVATION CENTERS—
THE UNIVERSITY OF UTAH

Introduction

Background

Early in the 1970s NSF was authorized by the U.S. Congress to undertake applied research to identify and test incentives to stimulate technological innovation. Part was conducted in the form of the Innovation Center Program experiment. This experiment involved the establishment of centers within universities to stimulate technological innovation and to increase the entrepreneurial tendencies of graduates as they pursued their careers. The experiment was designed to provide support for the Innovation Centers for five years, after which they would have to demonstrate their independent viability.

The Innovation Center Program experiment was established to address a wide range of barriers that exist within the innovation process. Some of the major barriers are the quality of available technology, the inability to match needs and capabilities, the inelastic nature of markets and supply systems, the size of the financial risks and rewards, the rigidity of institutional and intellectual patterns and the inability of available organizations to embark on sustained efforts to meet needs. To have an affect on these barriers, students participating in a center's activities are exposed to a combination of education, clinical applications, research and outside business/inventor assistance.

Each of the three centers currently in operation has interpreted somewhat differently the general charge of combining classroom activities with hands-on entrepreneurial coaching. The Innovation Center at M.I.T. focuses on the evaluation and maturation of ideas, whereas the center at Carnegie-Mellon University concentrates on technology transfer through entrepreneurship. The one at the University of Utah concentrates on providing inventor support and assistance. The following sections take a closer look at it.

The University of Utah Center was funded initially in 1977 by NSF. It acts as a service coordinator and arranges for services to be performed by various groups at the university. Figure 13 summarizes the range of participants and their roles in the Center's program.

Organization

During planning for the Utah Innovation Center, several models were considered, involving the establishment of the following:

1. a board of representatives from the College of Engineering, the College of Business, the University vice-president for research, the Center director and a community member;
2. a private, nonprofit corporation, separate from the university;
3. a center within the University of Utah Research Institute; and
4. a joint venture between the University's Colleges of Engineering and Business reporting to the respective deans, but with accounting functions administered through the normal academic channels with the Department of Mechanical Engineering.

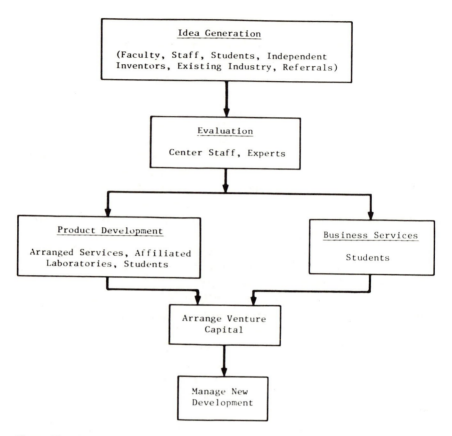

Figure 13.   Participants and their roles in the University of Utah Innovation Center.

This last mode of organization was selected because it would allow both colleges to share in the benefits and responsibilities of the program.

The Center's staff consists of four positions: a director, an associate director, an administrator and a secretary. The director oversees, and is responsible for, all programs and operations of the Center. The associate director coordinates the education program and acts as faculty liaison with the Center. The administrator, assisted by the secretary, is responsible for conducting the product development assistance program and coordinating the initial product screening and administration of subsequent evaluations. The administrator also acts as liaison with the University's agencies and laboratories and is responsible for financial administration of the grant from NSF.

Also, the Center supports a National Advisory Board made up of alumnists who are successful founders of technological enterprises. The members of this board serve as a primary resource group of guest lecturers for the innovation and entrepreneurship courses at the center. The board also serves to broaden contacts of the Center with industry and government while it assists in securing capital for center-sponsored new ventures.

Management Technique

Overview

This Innovation Center Program includes the following five basic program areas: (1) education, (2) innovation projects, (3) public affairs and communications, (4) new enterprise development, and (5) research on the innovation process.

The educational program consists of formal classes, "special topics" or arranged courses, short courses and seminars and technological creativity workshops. The business courses offered through the Center address such topics as the venturer's first year, goal setting, cash flow analysis, marketing, orientation and promotion of a new product, and new ventures feasibility. The courses presented by the College of Engineering include new innovations in the energy fields, innovation methods and entrepreneur career choice, writing proposals, patents and intellectual property and design methodology.

Much of the coursework conducted in the education program is directed toward the stimulation of new product or process ideas, testing the feasibility of new product ideas and development of business plans for the new products and processes.

The innovation projects sponsored by the Center are selected from ideas or proposals submitted by faculty, students, inventors in the community, industry and government agencies. The projects are evaluated by applying a set of selection criteria formulated to assess several aspects of the potential business and technological worth. A list of the project selection criteria [7] is presented below:

1. **High technology**. Does the project involve a significant technological component?
2. **Maturity of the project**. Has the innovator done enough work to demonstrate that he or she has thought the problem through, or does the inventor expect the Center to do it all?
3. **Novelty**. Is it really new or is it an obvious extrapolation of something already in existence?
4. **Needs**. Is the need for the product demonstrable or must applications be found?
5. **Evaluation**. Did the product receive favorable evaluation from those to whom it was demonstrated or explained?
6. **Safety**. Would pursuing the project within the Center impose hazards within the university setting? Would the product be safe in use?

7. **Scale**. Can the project be pursued within the limits of the Center's budget with a reasonable expectation of progress to a point at which it may be commercialized or additional funding may be secured?
8. **Resources**. Can the resources or services needed by the project be identified at the university?
9. **Project need**. Do the principals really need the Center's assistance or can they manage alone? Will the Center's services be significant in terms of the overall project?
10. **Timing**. Does the project show promise of conclusion within an acceptable period or is it too far into the future in terms of product development or business feasibility?
11. **Encumberance**. Is the project free to be presented to the Center?
12. **Bail-outs**. Is the product burdened by the financial and managerial problems of a firm in trouble?
13. **Personality**. Does the innovator appear to be reasonable and willing to take suggestions? Is he or she willing to work on the center's terms if accepted into its program?
14. **Equity**. Is the principal willing to share equity in the project in return for the Center's assistance?

The Innovation Center at the University of Utah has focused on selecting projects in fields in which the University can provide strong support by virtue of the areas of institutional prominence. The target technologies sponsored by the Center generally fall within the areas of bioengineering and biomedical products, computer and microcircuit technology, and energy technology. Some examples of projects under consideration by the Center are a professional development and training method for secondary school personnel, an improved machine for forming large cylindrical bales of hay and a mechanism that converts infrared radiation to the visible spectrum, thereby enhancing efficiency of photovoltaic cells.

The Center has employed different modes of new enterprise development. In some cases, a new development required little assistance from the Center as the prototype was developed and a product champion was carrying through with the management of the operation. In these cases, the Center provides guidance on an as-needed basis. In other circumstances, where management must be identified or developed, the Center may play a more active role in developing a work program or facilitating licensing of a new product or process to industry.

Selection and Implementation of the Management Technique

NSF began the Innovation Centers as an experiment to assess the ability of universities to nourish a practically oriented effort aimed at increasing American entrepreneurship and innovation. By providing a climate conducive to the development of new products and processes, it was anticipated that innovators naturally would take advantage of the climate. Each Innovation Center concentrated on one specific phase of the innovation process, although the entire process was always of concern.

The University of Utah's Innovation Center focused on providing services and technical (business and engineering) support to the entrepreneurs and innovators. A facility separate from the University was allocated for use by individuals whose projects have been evaluated and approved by the Center to provide an inexpensive sheltered environment in which to develop new products to the marketable stage.

Evaluation of the Management Technique

The Innovation Center Program experiment has been completed, but without long-term statistics it is difficult to assess fully the impact of these centers on the rate of innovation or level of entrepreneurship. Figures from three of the centers indicate that, over the five-hear period of the experiment more than 30 new ventures were developed successfully. Of these, 23 produced sales of more than $30 million, resulting in approximately 1000 new jobs. More than 2000 students have participated in these three centers' education programs, and approximately 2000 new product ideas were evaluated. Looking at the fourth year of the experiment, federal taxes collected as a result of the profits and wages attributed to the new products and processes generated by the Innovation Centers exceeded $2.5 million. This is approximately four times the annual federal investment in the Innovation Center Program.

The original plan for the Innovation Center experiment projected the eventual development of 40–50 centers located at universities around the country. The network of centers would not necessarily be financed solely by government funds. Based on the apparent success of the centers started in the Innovation Center Program, it is anticipated that centers will be initiated and financed by academic and, possibly, industrial communities.

There are several factors or characteristics of the centers that may have been influential in their successful operation. Because the Innovation Center Program was an experiment, there was as emphasis on testing hypotheses related to the education of entrepreneurs and on innovation. Innovation and entrepreneurship have been in the national spotlight, so the centers have benefited from a large and receptive audience. All the centers have been able to recount significant accomplishments, which has helped to attract recruits.

Other factors supporting the centers' successful development relate to their financing and innovations. NSF committed full funding to the centers for the entire experimental period. This permitted them to obtain institutional status and to avoid resource dissipation associated with renewal processes and accompanying uncertainties. The centers also served as a valuable liaison for new entrepreneurs and venture capitalists, thus providing an investment evaluation service to investors and a guide to sources of capital for innovators.

It has been said that the centers' success may not be evaluated accurately because some of the new ventures might have come into being without their help. Moreover, increased production and jobs may not be net gains due to the displacement of existing products. In addition, it is difficult to evaluate new product and process ideas in terms of appropriateness. It is anticipated, however, that centers established under the Innovation Center Program experiment will be self-sustaining and will continue with the program directives based on interest and support from the community, industry and the universities.

No other organizations employ this technique.

NATIONAL TECHNOLOGY FOUNDATION

Introduction

Background

The U.S. Congress has considered consolidating existing federal technology research and development programs within the proposed National Technology Foundation, which would serve as a resource that government officials could draw on to address long-range national scientific and technical problems. The foundation would attempt to address barriers that occur in the innovation process as a result of the lack of coordination of federally funded new technology development projects, the lack of foresight in planning scientific and engineering educational requirements, and the overlap of existing thrusts to establish university-industry relations.

The proposal to establish the National Technology Foundation stemmed from the concern of national political leaders, as well as from the industrial and scientific communities, about the decline in the nation's technological capabilities and level of innovation. In referring to innovation, the full structure of activities were seen to be affected, from creativity and invention on one end to the successful development of new products at the other. It was felt that this decline may be attributed to the overall slackened rate of development and application of technology, which can be attributed to the deteriorating relationship between government, industry, academia and labor. This malaise is affecting the nation's continued growth, and curing it will require integrating the goals and purposes of the professions with those of government, industry and academic institutions.

Other factors that call for a coordinated response from the institutions stimulating and conducting innovation in the U.S. are federal activities associated with regulation of capital, the fluctuating availability of venture capital, the change from a goods-oriented to a service-oriented economy, the increase in U.S. patents granted to foreign nationals and problems in engineering education.

Organization

An important consideration in the proposed establishment of the National Technology Foundation has been the role of the federal government. Because of the dispersed nature of the elements of private industry involved in the innovation process, the federal government must provide at least a coordinating role. As coordinator, the federal government can play the most valuable role in the areas of strategic planning, information gathering and overall evaluation. In terms of becoming more involved in the innovation process, however, the federal government cannot perform as well due to its limitations in market analysis and in determination of society's needs—skills quite well developed in the private sector.

Alternative organizational structures for the Foundation were considered, including (1) whether the responsibility of science and technology (or basic and applied research) should be separated within different agencies; (2) whether existing agencies could deal with the technological and innovation issues or whether a new organizational structure should be formed; and (3) the potential impacts of a new organization.

Must discussion was directed toward possible changes that would be made to the National Science Foundation or to the Department of Commerce, federal organizations that currently are directing and promoting innovation-related research or demonstration projects. A major argument against possible reorganization was the difference between basic and applied research. Existing organizations (particularly NSF) have organizational structures consisting of practices, procedures, protocols, power structures, philosophies, peer review groups, political constituencies and personnel that are not concerned with the support of technology development.

Management Technique

Overview

The National Technology Foundation would be established to provide an overall approach to innovation problems involving the use of human resources, enhanced capital formation and reduced regulatory burdens. The Foundation would take a multiprofessional approach to national problems involving various segments of society, including small businesses, management, universities and industry. It would identify emerging national needs related to taxes, capital formation, and technology development, all of which impact on innovation and productivity. Furthermore, the Foundation would provide a focal point for development and coordination of educational programs required to support the industries involved in technological innovation.

The National Technology Foundation, a new independent agency within the federal government, would, as proposed, perform the following primary activities:

1. Determine the relationships of technological developments and international technology transfers to the productivity, employment and world trade performance of the United States.
2. Develop improved indicators of the state of technology.
3. Determine the influence of economic conditions and government policies on industrial innovation and the development of technology.
4. Promote technology transfer from the federal government to private enterprise.
5. Support applied research in engineering and other disciplines as necessary in the national interest.

Many of these activities currently are being carried out by a diverse number of federal agencies involved in applied research and technology, including the National Bureau of Standards, the Patent and Trademark Office and the National Technical Information Service of the Department of Commerce, the Office of Small Business Research and Development, the Directorate for Engineering and Applied Sciences and several working groups of the National Science Foundation.

It is proposed that the Foundation should include an Office of the Director, Office of Small Business, Office of Institutional and Manpower Development, Office of Intergovernmental Technology, Office of Engineering and an Office of National Programs. A National Technology Board would be established to devise and power the policies and programs of the Foundation. The Board would be responsible for reporting to the Congress regarding policy issues affecting the Foundation.

Evaluation of the Management Technique

Legislation supporting the establishment of the National Technology Foundation has been countered with organizational or philosophical suggestions. One alternative approach is to establish a National Engineering Foundation, which would focus attention and resources on rehabilitating the research capacity of the engineering profession. Another alternative is the establishment of the national Professions, Technology and Engineering Foundation. It would be similar to the proposed National Technology Foundation, except that it would include the public professions (e.g., accounting) and provide an environment for national technical and professional resource allocations, as well as decision-making. Another suggested approach is the formation of an agency instead of a foundation. It would formulate policies to stimulate innovation and technology development and support engineering. The constituency also might include representation of labor, consumer, environmental and health interests.

REGULATIONS—HEALTH CARE INDUSTRY

Introduction

Background

Government regulation of the health care industry has had positive and negative impacts on technological innovation. Speaking in broad terms, government regulation refers to policies affecting licensure, reimbursement and investment in, or use of, new technology. Positively, through government funding of research and development of certain medical instrumentation devices and development of reimbursement structures to "reward" users of new technology, some new technologies such as the cat scanner have gained rapid acceptance. On the other hand, new technologies in the line of pharmaceuticals have been negatively impacted by regulation that controls the R&D process through standardization and codification.

This discussion focuses on the impact of regulations in the form of reimbursement by federal health care programs, such as Medicare and Medicaid, on the diffusion of new innovative health care technologies. Regulation of health care services reimbursement facilitates diffusion of new products by altering the incentives structure related to the use of new technologies. New medical technology has been characterized as expensive and a major contributor to rising health care costs. New medical technology also has been characterized as expensive and a major contributor to rising health care costs. In fact, new medical technology can be grouped into classes of large-scale medical technology, such as CAT scanners, renal analyzers, coronary bypass surgical equipment and fetal monitoring devices, and small-scale medical technology, such as testing devices. Although some observers of the American health economics system point to the introduction of large-scale medical technology as a major cause of rising health care costs, others argue that the higher medical costs may be attributed to the introduction of smaller technologies used for testing and monitoring. There is a growing concern that new medical technology should be evaluated more thoroughly to determine potential impacts on human health, as well as on health care economics.

Organization

Regulation of medical care technology by the federal government is accomplished quite easily through regulation of the policies of third-party health insurance companies, which are responsible for approximately 60% of payments for physicians' service and 90% of hospital services. If the services utilizing new technology are reimbursed by the third party system, the technology will tend to be adopted quickly. Unfortunately, this almost forced-use system encourages widespread acceptance of many new technologies before they are evaluated fully.

An example of the problem has been noted with the renal dialysis/renal transplantation program, which was quite inactive until 1972. Then, certain regulations extended Medicare coverage to individuals with chronic renal disease and who require hemodialysis or renal transplant. In 1978 approximately $15,000 per patient was expanded, a figure that will increase to a projected $46,000 by 1983. Questions have arisen regarding the amount of funds devoted to this disease, which is confined to 0.02% of the population. It has been suggested that medical technology advances be planned in terms of both the methods of funding research and development and those for introducing and evaluating new technologies.

Management Technique

The development of policies that impact on the introduction of new medical technology must involve the use of rewards or incentives for health care providers who purchase and use the technology. Several strategies have been offered to control costs of the development and use of technology, while stimulating the development of innovation in the medical technology industry. These approaches are outlined below:

1. Research and development funded by the federal government should demonstrate a definable link to a measurable health system goal. Unfortunately, many of the major successful medical technology contributions have resulted from unfocused, diverse research beginnings.
2. It has been suggested that new medical technology can be evaluated according to potential benefits that will accrue to society. The products with the highest priorities would then receive funding. Unfortunately, the health care industry does not lend itself to cost benefit analysis.
3. High-cost medical technology could be parceled to locations of greatest use, ensuring the greatest cost benefit; however, it is often difficult to determine the number of potential users.
4. Some movements have been initiated to abandon old technologies to make way financially for the new ones. Health care providers have not yet been able to establish a standard by which the old technologies should be measured.
5. Protocols could be developed to guide providers in the selection of appropriate (least expensive when possible) technologies for provision of reimbursable services.
6. Currently, medical specialists who use a high level of specialization technology are rewarded by the reimbursement system. A new system that does not necessarily reward specialist services could be installed. This would lessen the pressure to continually develop new medical speciality equipment and services.
7. An effort could be made to shift the fundamental values of technology users away from selected, high-cost diagnostic tools to a broad range of technology services.
8. Savings resulting from educated use of new technology should be passed on to the users of the technology in the form of increased prepayment rates!

Evaluation of the Management Technique

The most effective means of evaluating the impact of regulation on the diffusion of medical technology is to assess the cost benefits of one technology compared to another, which may or may not be affected by regulation. Effective regulatory mechanisms would realign incentives for using these technologies. Development of policies to create alternative, equitable reimbursement/incentives structures requires consideration of the short- and long-term costs to be borne by the users; the willingness of the system to accept the introduction of new technology; and the perceived benefits resulting from use of the technology. The information needed to derive this market assessment should be obtained from all segments of the health care industry, including technologists involved in the development of innovative medical technology, to ensure adequate support from all participants.

TECHNOLOGY TRANSFERS FROM GOVERNMENT LABORATORIES— ARGONNE NATIONAL LABORATORY

Introduction

Background

This management technique addresses the barriers that affect the marketing of new products developed in government laboratories. The technique was employed by Argonne National Laboratory and involves the participation of industry engineers and scientists in the laboratory setting in the technology development project. In 1979 Argonne was involved in the development of two projects using a lithium-selenium-based battery system, whose objectives were to develop batteries for U.S. Army vehicles and a power source medically compatible with the human anatomy. In 1973 the battery work shifted to address the development of a battery to be used with nuclear reactors to perform load-leveling for utilities. Recently, the focus of this battery research program concentrated on investigating batteries for electric vehicles.

Realizing the potential commercial applications of the lithium-iron sulfide battery, an industrial participation program was established at Argonne to assist in the development of a competitive, self-sustaining industry producing lithium-iron sulfide batteries to meet national needs with a minimum of federal support.

Implementation of this technology transfer mechanism, which utilizes industry representatives in a government laboratory, involves the steps shown in Figure 14. Through the use of this management technique, a link is formed between a technical resource, a government laboratory and industry that has strong resources and incentives to produce and market new products and processes.

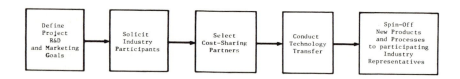

Figure 14.   Steps in implementing technology transfers from government.

Organization

This industrial participation program was conducted within the Battery Technology Development Group of Argonne National Laboratory. Argonne Lab is operated by the University of Chicago in cooperation with the Argonne Universities Association, a consortium of 30 mid-western universities. The largest single program effort there is concerned with the development of nuclear energy. Other major research efforts are being conducted in the fields of pressurized fluidized-bed combustion of coal, the development of magnetohydrodynamics for more efficient generation of electricity from coal and the development of a high-capacity lithium-iron sulfide battery. Argonne also has had major responsibility for assessing the biomedical and environmental effects of energy-related activities.

Approximately 20 professionals and 5 managers from Argonne, as well as 40 representatives of component suppliers or battery producers, were involved.

Management Technique

Overview

This management technique involves the interaction of industrial engineers, scientists and marketing personnel with their counterparts in a government laboratory. The goal of involving commercial representatives in the early phases of the innovation is to encourage the "transfer" of technology to the private sectors. Characteristics of this model of technology R&D are the assumption of risk by the government laboratory and the accessibility of information from the government regarding public policy and public welfare considerations. The costs are borne by both government and industrial participants, ensuring a follow-through commitment by the manufacturing and marketing firms.

In the actual research and development setting, the Argonne group leader oversees the work of the industrial participants. Even though there were representatives from competing firms participating at the same time, there was no competition to protect product information as the technology was in such early stages of development. The group leader must have a doctorate in chemical engineering. All industrial participants were technically oriented; no business representatives were included in the R&D phase of the battery systems development.

Selection and Implementation of
the Management Technique

The decision to involve industry representatives in the R&D program of battery systems was made to facilitate commercial development of the batteries, as well as to educate firms that eventually would become involved in battery production. The program was initiated by soliciting battery firms for the assignment of one or more of their engineers or scientists to Argonne's Battery Technology Development Program. These individuals would be responsible for assisting in the development work and assessing the technical and commercial potential of this type of battery. The solicitation was made through an advertisement in the *Commerce Business Daily.*

Five firms responded and sent personnel for periods of a few weeks to a year, starting in 1974. Usually no more than three industry representatives were in attendance at one time. Approximately 25 industrial participants from 12 firms have since participated in the program. In addition to the R&D activities, coordination among the marketing staffs of the industrial firms was encouraged, and in 1977 a market survey was conducted by Argonne to assist commercial businesses in assessing potential markets for the batteries.

During the spring and summer of 1975, Argonne initiated cell development and fabrication subcontracts with three of the battery firms that had industrial participants at Argonne. At the same time, Argonne was given management responsibility for a program being conducted by the Atomics International Division of Rockwell International, concerning the development of the lithium-iron sulfide batteries.

Since it subcontracted cell fabrication to the industrial partners, Argonne has fabricated internally only small research and engineering cells with unique features. More than 500 cells were produced by the industrial partners during 1979 alone. In the 1980 fiscal year, two subcontrators were developing multiplate cells with felt separators and 10-cell modules. In addition to their development and fabrication activities, both subcontractors plan to increase their capability for cell and module testing and posttest examinations of cells.

Requirements of cells for the stationary energy storage applications are being determined by Argonne and the industrial partners. The objectives of this effort are to extend the lithium-iron sulfide battery technology to meet the load-leveling battery application performance and cost goals; to develop load leveling cell and balance-of-plant design, which can demonstrate performance, life and cost objectives compatible with commercial system goals; and to demonstrate a load-leveling battery.

Evaluation of the Management Technique

Those involved in this industrial participation program have assumed that technology development planning can be done by industrial firms that expect to profit from the successful development, manufacture and marketing of the battery. To be successful, development of the lithium-iron sulfide battery requires a corporate commitment by firms involved in the development effort. For this reason, cost-sharing has been strongly encouraged and will be a requirement for future programs. A major benefit of cost-sharing by an industrial firm is that it permits the firm to develop proprietary positions that encourage commercialization.

In establishing and extending the contracts with the industrial partners, Argonne scientists and engineers have learned much about the problems of assembling batteries. As a result, they have redirected the development program. In return, the industrial firms have become better acquainted with restrictions imposed on cell and battery design by basic chemistry and materials considerations.

No other organization employs this technique.

SUBSIDIES—UTILITIES

Introduction

Background

The U.S. government has established a number of incentive mechanisms that may be utilized within or outside the regulatory process to stimulate innovation. These incentives may be used at several phases of innovation development—basic research, applied research, development and testing of a prototype design, demonstration of its technological and commercial feasibility, and diffusion. Incentives to stimulate innovation outside the regulatory process are subsidies, procurement protocols established by the government, patents and structured information transfer. Some incentive approaches used within the regulatory process include the employment of standards facilitating innovation, economic incentives and information transfer to aid decision-making, while removing regulatory uncertainty.

This discussion focuses on the use of subsidies to encourage and stimulate technological innovation within the private sector. Subsidies can be characterized by the transfer of income or something of value (such as supplies of goods or services) without the government's receiving something of equal value in return. The major reason why subsidies are offered to potential users of new technologies is to ease the financial risk associated with investment in an untested, unknown entity. Subsidies are used to overcome barriers associated with the diffusion of technology.

Management Technique

The federal government provides subsidies through several programs, not all of which are necessarily linked to innovation. Four broad categories of subsidies are: (1) direct cash; (2) tax; (3) credit; and (4) benefit- in-kind. Direct cash subsidies are cash payments made to individuals or firms that engage in a specific market activity, such as using specific technology or constructing a given facility. Tax subsidies, the largest group, reduce the tax liability of individuals or firms that engage in specific market activities. The investment tax credit awarded to firms is an example of a tax subsidy. Credit subsidies lower the rate of interest a private borrower pays for financing. Finally, benefit-in-kind subsidies are involved when the government sells goods or services below market value.

The government spurs innovation directly through cash subsidies for R&D. Subsidies also can affect the rate of technological acceptance by impacting on the relative prices of productive services and may be used to increase technological innovation indirectly. For example, an increased demand in an industry's output, stimulated as a result of subsidy, increases the demand for the goods of that industry's suppliers. The potential profits from supplying the subsidized industry make it possible either to develop a cost-reducing process or an improvement in the quality of supplies. The indirect impact of subsidies on the incentive of suppliers can have a great impact on such industries as electric power and the airlines, in which suppliers have conducted most of the research and developed most of the innovations.

Evaluation of the Management Technique

Based on an analysis of several case studies of subsidies used to stimulate innovation, it appears that demonstration programs used to test new technology may be more effective than subsidies, particularly when the risk of failure from using an untested technology is high relative to the benefits that could be realized by the innovator [8]. An example of this trade-off can be seen in the focus of federal assistance to first-generation nuclear power plants. Demonstration grants stimulated both the development and commercial introduction of the new technology.

In addition to stimulating the use of technology, which might not be utilized if a form of subsidy were not associated with its use, the increased output may induce more research and development. The subsidy programs generally were found to be more effective when accompanied by complementary government programs that have goals of development and commercial introduction of new technology. Often, however, the true impact of subsidies on the research, development and diffusion of new technology is difficult to measure due to the overlap of subsidy programs, as well as other government actions. To date, subsidy funding has had a positive, but largely unplanned, effect on technological change. However, it is seen as a viable tool for influencing the technological innovation process.

No other organization employs this technique.

REFERENCES

1. Mogee, M. E., and W. H. Schacht. "Industrial Innovation," Issue Brief No. IB80005, The Library of Congress Congressional Research Service, Major Issues System, Washington, DC (1980).
2. Clauser, H. R., Ed. *Progress in Assessing Technological Innovation* (Westport, CT: Technomic Publishing Co., Inc., 1974).
3. Rubenstein, A. H., A. K. Chakrabarti and R. D. O'Keefe. "Field Studies of the Technological Innovation Process," PB-253 403, Washington, DC (1974).
4. Gartner, J., and C. S. Naiman. "Making Technology Transfer Happen," *Res. Managemt.* 34–38 (May 1978).
5. DeCotiis, T. A., and L. Dyer. "Defining and Measuring Project Performance," *Res. Managemt.* (January 1979), pp. 17–22.
6. Chakrabarti, A. K., and A. H. Rubenstein. "Interorganizational Transfer of Technology: A Study of Adoption of NASA Innovations," *IEEE Trans. Eng. Managemt.* (February 1976), pp. 20–34.
7. Kapany, N. S., Ed. "Innovation Entrepreneurship and the University," Center for Innovation and Entrepreneurial Development, University of California, Santa Cruz (1978).
8. *Subsidies, Capital Formation, and Technological Change: Summary and Conclusions, Vol. 8,* (Cambridge, MA: Charles River Associates, Inc., 1978).

BIBLIOGRAPHY

Alkind, P. "Patent Office Fails as Mother of Invention," *The Washington Post*, Metro Section (August 18, 1980), p. 1.

American Management Association. "Effective Marketing through Franchises," speech presented to an orientation seminar on motivation of franchises.

Arbose, J. R. "Quality Control Circles: The West Adopts a Japanese Concept," *Int. Managemt.* (December 1980), pp. 31–39.

Bachman, R., Associate Director, University of Wisconsin Small Business Development Center, Madison, Wisconsin. Personal communication (June 17, 1981).

Bachman, R., Associate Director, University of Wisconsin Small Business Development Center, Madison, Wisconsin. Personal communication (June 29, 1981).

Beer, M., et al. "Performance Management System: Research Design, Introduction and Evaluation," *Personnel Psych.* (Autumn 1978), pp. 505–535.

Bivins, R., State Technology Applications Centers, National Aeronautics and Space Administration, Washington, DC. Personal communication (June 10, 1981).

Booz Allen Applied Research. "Government Procurement as an Incentive to Commercial Technology and Innovation," Bethesda, MD (1973).

Brown, W. S., Director, Innovation Centers at the University of Utah, Salt Lake City, Utah. Personal communication (June 26, 1981).

Burger, R. M. "An Analysis of the National Science Foundation's Innovation Centers Experiment," National Science Foundation, Washington, DC (1978).

Chakrabarti, A. K. "The Role of Champion in Product Innovation," *Cal. Managemt. Rev.* (Winter 1974), pp. 58–63.

Chilenskas, A. A., LI/MS Battery Program, Manager of Battery Technology Development, Argonne National Laboratory, Argonne, Illinois. Personal communication (August 18, 1981).

Clogston, A. M., Director of Physics Research, Bell Laboratories, Murray Hill, New Jersey, Personal communication (August 26, 1981).

Colton, R. M. "Technological Innovation Through Entrepreneurship," *Eng. Ed.* (November 1978), pp. 193–197.

143

Colton, R. M. "National Science Foundation Experience with University-Industry Centers for Scientific Research and Technological Innovation: An Analysis of Issues, Characteristics and Criteria for their Establishment," National Science Foundation, Washington, DC (1980).

Colton, R. M. "The NSF Innovation Center Experiment," *Mech. Eng.* (August 1980), pp. 26–27.

Colton, R. M., Project Director, Innovation Center Program, National Science Foundation, Washington, DC. Personal communication (April 9/ August 13, 1981).

Comer, J. M., R. D. O'Keefe and A. A. Chilenskas. "Technology Transfer from Government Laboratories to Industrial Markets," *Ind. Res. Managemt.* (September 1980), pp. 63–67.

DeCotiis, T. A., Professor, University of South Carolina, Columbia, SC. Personal communication (August 1981).

Deming, W. E. "What Top Management Must Do," in "Japan: Quality Control and Innovation," *Bus. Week* (July 20, 1981), pp. 19–20.

Denver Research Institute. "Applications of Aerospace Technology in the Electric Power Industry," University of Denver, Denver, CO (1973).

Donaldson, Lufkin & Jenrette, Inc. "Donaldson, Lufkin & Jenrette, Inc. Annual Report," New York (1980).

Ellinghaus, W. M. "Impact of Industrial Innovation in the 1980's—The Telecommunications/ Electronics/ Computer Industry," *Res. Managemt.* (March 1981), pp. 12–14.

Ellis, L. W. "Effective Use of Temporary Groups for New Product Development," *Res. Managemt* (January 1979), pp. 31–34.

Ellis, L. W., Vice-President, Engineering, Bristol Babcock, Waterbury, Connecticut. Personal communication (July 10, 1981).

Fast, N. D. "New Venture Departments: Organizing for Innovation," *Ind. Marketing Managemt.* (July 1978), pp. 77–88.

Fast, N. D. "A Visit to the New Venture Graveyard," *Res. Managemt.* (March 1979), pp. 18–22.

Fast, N. D. "Pitfalls of Corporate Venturing," *Res. Managemt.* (March 1981), pp. 21–24.

Ford Motor Company. "Annual Report, 1980," Dearborn, MI (1981).

Ford, D., and C. Ryan "Taking Technology to Market," *Harvard Bus. Rev.* (March/ April 1981), pp. 117–126.

Geschka, H. "Introduction and Use of Idea-Generating Methods," *Res. Managemt.* (May 1978), pp. 25–28.

Globe, S., A. W. Levy and C. M. Schwartz. "Key Factors and Events in the Innovation Process," *Res. Managemt.* (July 1973), pp. 8–15.

Goodhart, M. American Hospital Association, Chicago, Illinois. Personal communication (July 22, 1981).

B. F. Goodrich Company. "B. F. Goodrich Annual Report," Akron, OH (1980).

Greenstein, H. "Licensing New Product Technology," *Ind. R/D* (June 1978), pp. 122–127.

Gruber, W. H., and D. G. Marquis, Eds. *Factors in the Transfer of Technology.* (Cambridge, MA: The M.I.T. Press, 1969).

Haggarty, J. J. "Spinoff 1980: An Annual Report," Office of Space and Terrestrial Applications, Technology Transfer Division, National Aeronautics and Space Administration, Washington, DC (1980).

Hannay, N. B. "Industrial Innovation and Federal Policy," Bell Laboratories, Murray Hill, NJ (1979).

Hayes, R., Manager of Technology Investment Planning, Xerox Corporation, Stamford, Connecticut. Personal communication (July 9, 1981).

Hill, C. T., and J. T. Utterback. "The Dynamics of Product and Process Innovation," *Managemt. Rev.* (January 1980), pp. 14–20.

Hill, R. M., and J. D. Hlavacek. "Learning from Failure: Ten Guidelines for Venture Management," *Calif. Managemt. Rev.* (Summer 1977), pp. 5–16.

Hodges, P. "The Invention Factory," *Output* (1981), pp. 48–57.

Hopkins, D. S. "The Roles of Project Teams and Venture Groups in New Product Development," *Res. Managemt.* (January 1975), pp. 7–12.

Howe, R. J., et al. "Introducing Innovation Through O.D.," *Managemt. Rev.* (February 1978), pp. 52–56.

Imhoff, M. and P. Lenon. *Wisconsin Business Resource Directory*, (Madison, WI: University of Wisconsin Small Business Development Center, Madison, WI 1980).

International Harvester Company. "Corporate Policy Statement on Patents," Chicago, IL (March 1976).

International Harvester Company. "I-H Professional Compensation Programs," Chicago, IL (1980).

"ITT: Groping for a New Strategy," *Bus. Week* (December 15, 1980), pp. 66–69.

"ITT Terrific?" *Forbes* (December 22, 1980), pp. 62.

Kaufman, J. "How NASA Helped Industry," *Wall Street J.* (August 28, 1981).

Kohorn, Z. "New England Energy Development Systems Center (NEEDS), An Experiment in University-Industry Cooperative Research," MTR-79W00188. Report to the National Science Foundation, Metrek Division of the MITRE Corporation, McLean, VA (1979).

Laubach, G. D. "The Pharmaceutical/Health Industry," *Res. Managemt.* (March 1981), pp. 9–11.

Levitt, T. "Production-line Approach to Service," *Harvard Bus. Rev.* September–October 1972), pp. 14–16.

McCamus, D. R. "Managing Xerox for Technological Change," *Conf. Bd Record* (August 1975), pp. 32–34.

McClenahen, J. S. "Bringing Home Japan's Lessons," *Industry Week* (February 23, 1981), pp. 69–73.

McDonald's Corporation. "McDonald's Twenty-Five Years," Oakbrook, IL (1980).

McDonald's Corporation. "McDonald's Corporation, 1980 Annual Report," Oakbrook, IL (1981).

McGuire, E. "Generating New Product Ideas, The Conference Board, Ontario" (1972).

McKay, K. G. "The Bell System: Innovation through Interaction," Bell Laboratories, Murray Hill, NJ (1975).

McKay, K. G. "Innovation: A Bell System Commitment," Bell Laboratories, Murray Hill, NJ (1978).

McManus, G. "Attack on Patent System Threatens Innovation," *Iron Age* (January 19, 1976), pp. 34–35.

"Medical Technological Advances and Health Care Costs," *J. Med. Soc. NJ* (January 1980), pp. 13–15.

Milliken, J. G. "Management Contributions of Space Technology: An Analytical Report," Denver Research Institute, Denver, CO (March 17, 1969).

Minnesota Mining and Manufacturing Company. "3M 1980 Annual Report," St. Paul, MN (1980).

The MITRE Corporation. "The 1980 Annual Report of the MITRE Corporation," Bedford, MA (1980).

The MITRE Corporation. "Twenty Years of Change: A MITRE Retrospective," Bedford, MA (1978).

Moloney, T. W., and D. E. Rogers. "Medical Technology—A Different View of the Contentious Debate over Costs," *New England Med. J.* (December 27, 1979), pp. 1413–1419.

Moreland, D., Staff Director of Licensing, McDonald's Corporation, Oak Brook, IL. Personal communication (August 9, 1981).

Morrison M. "The Venture Capitalist Who Tries to Win Them All," *Fortune* (January 28, 1980), pp. 96–99.

"NSF Under Challenge from Congress, Engineers," *Science* (September 26, 1980).

Oren, S., M. H. Rothkopf and R. D. Smallwood. "Evaluating a New Market: a Forecasting System for Nonimpact Computer Printers," *Interfaces* (December 1980), pp. 76–87.

Parker, S. K., Industry Branch, Solar Energy Research Institute, Golden, Colorado. Personal communication (June 24, 1981).

Potter, R. J. "I-H Sets Plan to Reward Innovation," *Ind. Res. Develop.* (June 1980), pp. 108–110.

Potter, R. J. "Professional Compensation Program," *Up Front Agric. Machinery* (Fall 1980), pp. 16–17.

Potter, R. J. "Motivate and Compensate Individual Professional Contributors," Unpublished results (1981).

Potter, R. J., Senior Vice President and Chief Technical Officer, International Harvester Company, Chicago, IL. Personal communication (August 19, 1981).

Props, J. K., National Licensing Director, McDonald's Corporation, Oak Brook, IL. Personal communication (August 9, 1981).

Reeder, L., Senior Vice-President, Development Banking, Donaldson, Lufkin & Jenrette, Inc., New York. Personal communication (August 26, 1981).

Rendall, E. "Quality Circles—A 'Third Wave' Intervention," *Training Develop. J.* August, (August 1980), pp. 28–31.

Roark, M., Information Systems, Ford Motor Company. Personal communication (June 18, 1981).

Robbins, M. D., Director, Colorado Energy Research Institute, Lakewood Colorado. Personal communication (June 25,1981).

Ross, I. M. "The Impact of Innovation," *NJ Bell J.* (Winter 1980/81), pp. 9–14.

Simmons, D. L., Organizational Analyst, B.F. Goodrich Company, Akron, OH. Personal communication (March 26, 1981).

Smith, L. "The Lures and Limits of Innovation," *Fortune* (October 20, 1980), pp. 84–94.

Sounder, W. E. "Encouraging Entrepreneurship in the Large Corporations," *Res. Managemt.* (May 1981), pp. 18–22.

"Special Report: The New Industrial Relations," *Bus. Week* (May 11, 1981), 84–96.

Steel, L. W. *Innovation in Big Business.* (New York: Elsevier North-Holland, Inc., 1975).

Steward, J., Employee Involvement Program, Ford Motor Company, Dearborn, Michigan. Personal communication (June 18, 1981).

"The Tale of the Tape—50 Years of Innovation at 3M," *The Office* (September 1980), pp. 101–104.

Tassey, G., National Bureau of Standards, Washington, DC. Personal communication (March 27, 1981).

Taxonomy of Incentive Approaches for Stimulating Innovation (Washington, D.C.: Public Interest Economics Center, 1978).

"Technology Costs and Evaluation," *New England J.* (December 27, 1979), pp. 1444.

Thornton, R. "The Patent Issue—Effects of the Patent System on Innovation," *Res. Managemt.* (March 1979), pp. 33–35.

Tilton, J. E. "International Diffusion of Technology: The Case of Semiconductors," The Brookings Institution, Washington, DC (1971).

Tuhy, C. "The (Sometimes) Rousing Rewards of the Lone Inventor," *Money* (January 1981), pp. 51–52.

UAW—Ford National Joint Committee on Employee Involvement. "Employee Involvement: A Handbook on the UAW-Ford Process for Local Unions and Management" (1980).

U.S. House of Representatives. "An Analysis of the Hearings on H.R. 6910, The National Technology Foundation Act of 1980," Subcommittee on Science Research and Technology, Committee on Science and Technology, Washington, DC (May 1981).

Udell, G. G. "Stimulating Innovation through Management and Technical Assistance," Wisconsin Innovation Service Center, University of Wisconsin Small Business Development Center, Madison, WI (1980).

Udell, G. G., and K. G. Baker. *PIES-II Manual for Innovation Evaluation.* Wisconsin Innovation Service Center, University of Wisconsin Small Business Development Center, Madison, WI (1980).

Vesper, K. H., and G. H. Thomas. "How Venture Management Fares in Innovative Companies," *Res. Managemt.* (May 1973), pp. 30–32.

Wagner, J. L. "Reimbursement Shapes Market for Technology." *Hospitals* (June 1, 1979), pp. 91–94.

Wagner, J. Urban Institute, Washington, DC. Personal communication (August 11, 1981).

Wagner, J. L., and M. J. Krieger. "The Price of Progress? Medical Technology and Health Care Costs," *J. Contemporary Bus.* (April 1980), pp. 19–33.

Walcoff, C. A. "Planning and Management of Projects and Exploration of Results: Phase I Final Report," MTR-80W315, The MITRE Corporation, McLean, VA (1980).

Wilemon, D. L., and G. R. Germill. "The Venture Manager as a Corporate Innovator," *Calif. Managemt Rev.* (Fall 1973), pp. 49–56.

Wilson, "The Effect of Technological Environment and Product Rivalry on R&D Effort and Licensing of Inventions," *Rev. Econ. Stat.*, (May 1976), pp. 171–178..

Wright, P. "Government Efforts to Facilitate Technical Transfer: The NASA Experience," in *Factors in the Transfer of Technology*, W. H. Gruber and D. G. Marquis, Eds. (Cambridge, MA: The M.I.T. Press, 1969).

Xerox Corporation. "Xerox Corporation 1980 Annual Report," Stamford, CT (1980).

Xerox Corporation. "Xerox Research Centers," Stamford, CT (1980).

Yager, E. "Quality Circle: A Tool for the '80," *Training Develop. J.* (August 1980), pp. 60–62.

INDEX